CCTV

CCTV
A Technology Under the Radar?

INGA KROENER
Trilateral Research and Consulting, UK

LONDON AND NEW YORK

First published 2014 by Ashgate Publishing

2 Park Square, Milton Park, Abingdon, Oxfordshire OX14 4RN
52 Vanderbilt Avenue, New York, NY 10017

Routledge is an imprint of the Taylor & Francis Group, an informa business

First issued in paperback 2020

British Library Cataloguing in Publication Data
A catalogue record for this book is available from the British Library

The Library of Congress has cataloged the printed edition as follows:
Kroener, Inga.
CCTV : a technology under the radar? / by Inga Kroener.
pages cm
Includes bibliographical references and index.
ISBN 978-1-4094-2345-4 (hardback)
1. Crime prevention--Social aspects--Great Britain--History.
2. Video surveillance--Great Britain--History. 3. National security--Great Britain--History. 4. Privacy, Right of--Great Britain--History. I. Title.

HV7434.G7.K76 2014
364.4--dc23

2013045479

ISBN 978-1-4094-2345-4 (hbk)
ISBN 978-0-367-60068-6 (pbk)

Contents

Acknowledgements

I owe a huge thanks to a number of people for their help, advice and proof reading of various chapters: Jon Agar, Chris Williams, Chris Newstead, Kieron Flanagan, Lisa Southern, and Daniel Neyland. Thank you so much for giving up your time. A big thank you as well to Claire Jarvis and Gemma Hayman at Ashgate for their help throughout this book writing process.

To all the people who participated in the questionnaire and interviews – thank you.

And to my friends and family – thank you for your continuous encouragement throughout this process. I also need to include a special thank you to James for his unwavering support.

This book is dedicated to my parents. For everything you have done for me.

List of Abbreviations

ACPO	Association of Chief Police Officers
ANPR	Automatic Number Plate Recognition
CCTV	Closed circuit television
DHS	Department of Homeland Security
DPA	Data Protection Act 1998
Nacro	National Association for the Care and Resettlement of Offenders
RIPA	Regulation of Investigatory Powers Act 2000
STS	Science and Technology Studies

Introduction

In 1991, in a seminal paper published in *Scientific American*, Mark Weiser made the following statement: 'the most profound technologies are those that disappear. They weave themselves into the fabric of everyday life until they are almost indistinguishable from it'. He was referring to computers, but perhaps this statement resonates particularly with CCTV and its widespread use in Britain. Why has Britain become so camera-surveilled? What were the factors involved in its development and subsequent use as a technology to protect people from crime? How has a seemingly mundane technology become such an important tool for politicians and the police? And how did the widespread deployment of a surveillance technology occur seemingly under the radar of the public?[1]

A number of academics and other commentators have raised concerns over Britain's entry into a 'surveillance society' in recent years (see for example, Lyon 2001). There has undoubtedly been an increase in the amount of information collected, stored and processed on the British population over recent years, but whether this means we now live in a surveillance society is debatable. Surveillance practices and technologies certainly form a part of our society, but the line over which we theoretically cross into a surveillance society is an arbitrary one.

Nonetheless, this increase in surveillance activities, technologies, and actions has prompted a burgeoning literature focusing on power relations (Lyon 2007), social sorting (Lyon 2003), classification (Haggerty and Ericson 2006), social order (Rule 1973), the 'Panopticon' (Poster 1990),[2] and networks and categories (Marx 2002), to name a few aspects. What this literature tells us is that surveillance is a complex issue, tied together with notions of identity, discipline, power, social sorting, and categorisation. It also shows us that there are a variety of approaches to studying surveillance. In all of these studies there is an underlying message

1 Although the term 'the public' has an established cultural currency, it has been widely acknowledged for decades that this is a constructed or catch all term for a diverse set of publics, or shifting set of public groups who coalesce around specific issues, or who are manufactured through surveys and opinion polling (see for example, Lippman 1922; Dewey 1927). In this book I use the term the public to refer to the many different public groups who occupy the public sphere at any given moment.

2 The Panopticon was an institutional building designed by the philosopher and social theorist Jeremy Bentham in the late eighteenth century. In 1975 the French philosopher Michel drew on Bentham's idea to argue that a range of other social institutions had come to resemble panopticons in terms of the exercise of power and vision. The Panopticon has since been used by a range of surveillance studies theorists as a metaphor for analysing contemporary surveillance.

that surveillance is widespread and endemic. However, as Rule has argued, surveillance practices are also constrained in terms of their size and the amount of information they can process. They are not absolutes. The existence of a variety of surveillance systems and practices does not necessarily herald our entry into a surveillance society.

CCTV

In the context of thinking about CCTV (which is the main focus of this book) this notion that we have entered into a surveillance society becomes an important one. If CCTV is assumed to simply be part of this surveillance society then there is little left to question in terms of its history. It becomes part of an inevitable evolution of a society into one that is based primarily on surveillance practices. Even the popular claim that there are over 4.2 million CCTV cameras in Britain (which, although disputed, still appears in some recent academic literature and media reports) implies that there is nothing left to resist or question. There is a question that needs to be addressed, however. Why has Britain become so camera-surveilled?

There is a growing literature concerned with CCTV. CCTV is heralded as expanding disciplinary social control as well as targeting individuals in terms of suspicious behaviour (see for example, Norris 2003). Risk and security have also become important themes within research. Norris and Armstrong (1999) argue that systems of surveillance are ushered in by motifs of the 'stranger society' in relation to a loss of communication and community. This loss of communication leads to less knowledge of others' actions and a reduced ability to predict others' behaviour, in turn leading to an increased sense of risk and fear. Camera systems are then used to manage these feelings of risk and fear, although it can be argued that they may actually have the opposite effect (Kroener and Neyland 2012).

Other literatures concentrate on public space and the impact of CCTV on this space, often bound with issues related to social control, including the Panopticon, exclusion, discrimination and consumption. The use of CCTV in public space is discussed in relation to a fortification of the city, which arises from a sense of danger and feelings of fear, developed through interactions with strangers (Fyfe, Bannister, and Kearns 1998). These new forms of social interaction are also said to have led to new forms of social control where the city has become a place to manage out inappropriate behaviour in the new territories of consumption (McCahill 1998). Some now suggest that civil space has become consumer space, with a control of public space in the town centre where certain types of behaviour are not accepted (Reeve 1998). These literatures all suggest that cities have become Panopticons through the use of CCTV (Ainley 1998; Cohen 1985; Davis 1990; Lyon 1994; Oc and Tiesdell 1997). Furthermore, Koskela (2000) suggests that video surveillance changes the way power is exercised and modifies emotional experiences in urban spaces, as well as affecting the way in which reality is conceptualised and understood. In this sense then, Koskela (2000, p.243) argues that CCTV and surveillance 'contributes to the production of urban space'.

Fyfe and Bannister (1998) look at the economic and political restructuring of urban space, concentrating on the idea of the 'fortress impulse' in contemporary urban design. They highlight that there are numerous, complex, economic and political forces behind the rise of CCTV, which in turn means a change in social experience for those who occupy these spaces. The authors argue that those in charge of city centres and shopping areas install CCTV in a bid to emulate (for the public) the perceived safety and security of out-of-town retail parks and shopping centres. The use of CCTV systems therefore extends further than dealing with crime or the fear of crime, stating that they are used for dealing with inappropriate behaviour, suspicious youths and so on. This has the effect of privatising public space, enabling commercial goals for who controls what and who does and does not belong there. They conclude that research on CCTV to date has primarily focused on evaluating the before and after impacts of cameras on the level and distribution of crime and, whilst important, they state that this 'research agenda has marginalised the broader economic, political and socio-cultural issues surrounding the development of CCTV surveillance in public space'.

These literatures suggest that public space has changed substantially with the introduction of CCTV systems. They are useful when thinking about the discourse justifying the use of CCTV in city centres and other public space. I will return to this in more detail in Chapter 5.

Over the last 25 years, there have also been a number of studies focusing on issues of effectiveness, including public awareness and attitudes towards CCTV systems (see for example, Honess and Charman 1992; Bulos and Sarno 1994; Welsh and Farrington 2002; Tilley 1998). There has also been an interest in ascertaining levels of public acceptance of CCTV (including Williams and Johnstone 2000; Gill, Bryan and Allen 2007; Spriggs 2005). These literatures raise an important question. A number of studies have shown CCTV to be ineffective for the purposes for which it has been installed (to prevent crime, for instance). An important question then arises – if CCTV does not prevent crime then how has it been (and still is) constructed as a technology that provides safety and security? What are the influences that have shaped CCTV to become viewed as a surveillance and security technology? How is the myth of CCTV promoted and perpetuated? These are all key questions I address in the book.

The history of CCTV has primarily been looked at from the 1990s onwards. Commentators tend to dismiss any earlier history as inconsequential, marking the beginning of the 1990s as the start of the CCTV 'trend'. There is some work in the area of CCTV and policing during the 1960s and 1970s (see, Williams 2003), however this focuses only on one aspect of the use of CCTV. Although the use of CCTV increased dramatically from the 1990s onwards, it was not just an injection of funding or a few important moments (such as the James Bulger case) that set its later trajectory of widespread use, but rather a longer history of changes in policing, criminal justice, public opinion, social change, and governance, and a variety of voices and influences that shaped and constructed CCTV into a surveillance technology.

In the remainder of this introduction, I introduce my methodology and theoretical framework that will underpin my analysis of why Britain has become so camera-surveilled. I then set out the structure of this book.

Methodology

In order to answer the main question that I pose in this book – why has Britain become so camera-surveilled? – I draw on a Science and Technology Studies (STS) perspective. This allows the adoption of an interpretive framework, focusing on socio- technical and politico-economic structures, dynamics and histories.

Underlying this question is a subsidiary question focusing on the public in relation to CCTV. The peculiar role of the public in apparently accepting and not contesting CCTV in a time of supposed resistance to new technologies (such as biotechnology and nuclear power) is an issue that seems worth exploring (and one that is missing from the academic literature on CCTV). There are studies that explore public acceptance of CCTV systems,[3] but what is missing from the literature is an exploration of how CCTV has been presented to the public. I undertake this exploration through an analysis of policy documentation and a media analysis concentrating on the period 1992–2008.

There are a wide range of approaches to understanding technology and society, and the social contexts within which it is used. There are those explanations that favour the role of technology in determining society; those that favour the role of society in shaping or determining technology; and those that favour neither. In the following section, I will go into more detail about each of these perspectives.

Under the first perspective, known as technological determinism, technology is seen as an autonomous force, determining and structuring society. It determines social change. Social forces (such as, for example, politics) only start to play a part when the technology has already been introduced into society, i.e. they have no part to play in the development or definition of the technology (Bimber 1994). For example, Jacques Ellul (1962, p.10) describes technology (or 'technique' as he terms it at the time) as being 'artificial, autonomous, self-determining, and independent of all human intervention'.

In contrast, other commentators argue that technologies are inherently social and that they cannot be abstracted from the social context. This forms the basis for the second perspective, that of social constructionism. Under this perspective, it is society that becomes the influencing factor where technologies are shaped and constructed by the social context in which they are developed. Society shapes technology, and technology has social relations imbedded into it throughout its entire lifecycle (see for example, Wajcman 1991). The social constructionist

3 I cover these literatures in Chapter 5.

perspective does however also hold some danger. There is a risk at the extreme end of the constructionist perspective that it veers into a territory of social determinism. The 'thingness' of things (Heidegger 1977) can become forgotten. Technologies under this perspective become purely social.

Under the third perspective, both technologies and the social have agency and can have effects and influences. This perspective is known as co-construction (Hughes 1994) or co-shaping (Verbeek 2005), and this is the perspective that I follow in this book. I believe that what is missing from explanations of the rise of CCTV to date is a theoretical underpinning that this technology and its uses are not and have not been pre-determined. A variety of influences, social, political and economic, have influenced its design, development, and its very definition but technology has also had an effect on the social context in which it has developed.

Gaining knowledge of any technology or technological system is problematic and complex. To differentiate and distinguish between what is 'social' and 'technical' is arguably impossible. In the case of CCTV, it is not simply a technology that makes up the system but a network of actors (the police, CCTV operatives, the public, and so on) that also engage with and define the technology in the context in which it is utilised and put into practice. The camera itself does not stand alone as a system. The camera also does not sit outside of society. It becomes constructed and defined in terms of its applications and uses. In turn, its applications and uses also influence how people view the system and what it then eventually becomes used for (in the case of CCTV, the technology is first used as a transmission technology, only eventually becoming a surveillance technology many years later). Its meaning is 'flexible' (Bijker 1995).

Pfaffenberger (1992) argues that politics are constructed through technological means. This, he suggest, happens through a process of 'affordances', by which perceived properties of technologies (or artifacts) suggest how they should be used, or responded to. Affordances are social constructions and depend on a certain 'reading' of uses or meanings into that technology. He goes on to suggest that in order for a particular reading (affordance) of a technology to become dominant, it also needs to be 'discursively regulated'. This means that the reading or affordance needs to be justified and legitimated by a 'sufficiently persuasive discourse' (Brey 2005, p.68). Using this notion as a theoretical basis for understanding CCTV is a useful one. What is the underlying political discourse that has justified the use of CCTV as a surveillance technology?

There have always been competing versions of CCTV and this will be the focus of this book. The main question of why Britain has become so camera-surveilled will be answered through an analysis of the uses of CCTV, its potential uses, discourse surrounding the technology, the wants and needs of politicians and the police in relation to the technology (what did this technology need to provide in order to be used?), the role of the public, regulation, and the influence of the media and popular culture.

I will argue that CCTV was not inevitably a surveillance technology but it has come to be defined as a technology that provides security, and that this adds to the general thesis of inevitability that resides in surveillance studies literature. CCTV is promoted and justified politically in this context as well.

The Structure of the Book

Chapter 1 focuses on surveillance in general terms. It looks at surveillance and how it has been theorised, whether it is something new, and how thinking about developments in surveillance more generally can be useful when thinking about CCTV in particular.

Chapter 2 provides a socio-historical context to the remainder of the discussion in the book. It focuses on changes in society, politics and the economy in post-war Britain (since the 1950s). Furthermore, it looks at developments in science and technology, information and communication technologies, and science policy. It also begins to look at ideas of the public, and public engagement with science and technology (these ideas are picked up once again in Chapters 5 and 6). The aim of this chapter is to provide a wide lens through which to view the development of CCTV in Britain; to begin to see the technology as part of a set of complex social, political and cultural relationships between a variety of actors and influences.

Chapter 3 explores developments in the area of criminal justice and policing in post-war Britain. Chapters 1 and 2 provide a background of changes in surveillance technologies and practices, and the socio-historical context in which these changes have taken place in Britain. Chapter 3 begins to look in more detail at changes in criminal justice, policing, and police use of CCTV from the 1950s. It highlights important shifts in approaches to law and order, and the policing of public space, and how these changes opened up a space in which CCTV could begin to be used. It also highlights that, although for some CCTV is seen as a tool for policing, that there was in fact a general reluctance on the part of the police to use the technology for purposes other than traffic control. Looking at developments in approaches to law and order and criminal justice provides an alternative explanation of why CCTV began to be utilised more widely – that it was changes in political approaches to tackling crime that pushed the technology onto centre stage. This idea is developed further in Chapter 4.

Chapter 4 develops a more general history of CCTV from the 1950s. In light of the developments in police uses of CCTV covered in chapter 3, this chapter analyses the use of CCTV in a range of situations and contexts. The chapter tells a story of a technology that was initially used in a range of settings: as a transmission technology (in both the transport and medical sectors); as an 'extra eye' for health and safety evaluations; and a technology used by politicians to improve the quality of political broadcasts. This chapter shows the gradual shift from a mundane technology used for a variety of purposes and in a variety of settings, to one firmly rooted politically and socially as a surveillance technology (the use of which is justified in terms of providing security and safety). Also drawing on the background provided in Chapters 2 and 3, this chapter formulates the argument that the use of CCTV for the security and surveillance purposes for which it is currently used was not inevitable. The history that is drawn together and developed in Chapters 2–4 shows that there were a range of potential capabilities and uses for CCTV. Its development into a tool for surveillance was not pre-determined (this was not one of the first uses for which it was developed). These three chapters show the history of the technology in light of a wide socio-political analysis, highlighting the variety of actors and influences involved in its evolution into a political tool of surveillance.

Chapter 5 focuses on the public aspects of CCTV, with specific reference to political discourse surrounding the technology (at both local and national level), as well as providing the results of a media analysis focusing on the representation of CCTV by the print media in Britain during the period 1992–2008. This section also provides the results of a small-scale study undertaken on two local authority run estates in London. The project studied was a joint undertaking by a local regeneration agency, a national government department, the European Union, and the private sector, to bring CCTV and broadband to two estates in London. Although this study only had a small number of participants, and is therefore not representative, it highlights an important aspect of the relationship between CCTV, politicians and the public. The public are viewed as passive figures and are not consulted (even though local knowledge in this instance could have potentially saved a lot of public money). The technology is firmly rooted outside of any socio-cultural framework (it is depicted as the solution to crime on the estates), and politically the project carries weight as it brings a technological fix to two estates with a fairly low socio-economic status (it is part of a wider initiative to bridge the so-called 'digital divide').

Chapter 6 focuses on the international context. The first part of the chapter focuses on the regulation of CCTV in Britain, showing that CCTV has enjoyed a privileged position in the past – a time of being almost entirely unregulated. This situation has now changed. There is a range of legislation in existence that is applicable to video surveillance. This section of the chapter also discusses to what extent regulation in the area of CCTV has been enforced. The second part of the chapter analyses the international regulation, use and acceptance of video surveillance. The aim of this book is to provide an explanation of why Britain has so many CCTV cameras. One of the most interesting things about CCTV in Britain is that it has become so much more widespread and utilised here than anywhere else (with the possible exception of China). What are the social, cultural, legal and political experiences of other countries? What are the factors or experiences that have hindered the rise of CCTV in other countries?

Chapter 1
Surveillance

Introduction

In a recent report from the Surveillance Studies Network for the Information Commissioner's Office (2006) it was argued that:

> We live in a surveillance society. It is pointless to talk about surveillance society in the future tense. In all the rich countries of the world everyday life is suffused with surveillance encounters, not merely from dawn to dusk but 24/7. Some encounters obtrude into the routine, like when we get a ticket for running a red light when no one was around but the camera. But the majority are now just part of the fabric of daily life.

We are undoubtedly surveilled in everyday life. However, it is a mistake to assume that all encounters with government run technology are also surveillance encounters. Technologies (even those with potential surveillance capabilities) inhabit a variety of different roles and definitions. In the above example, a red light camera is cited as being a surveillance technology. However, it could also be argued that it is primarily a road safety technology. Information is not being collected and processed in order to surveille the population generally, but to target the specific instance of driving through a red light. Although within the surveillance studies community the subtleties of taking this position are often carefully worked out (technologies are also perceived as mundane, ordinary, and with a variety of uses), this argument that we have entered a surveillance society often finds its way into public discourse in far simpler terms.

Claims in the academic sphere that we have entered a surveillance society are not always wholly negative. However, in terms of public discourse, statements that we are living in a surveillance society are often cited as an absolute. In the chapters that follow, I build a history of CCTV that eventually leads to an argument for greater public engagement in relation to CCTV (and surveillance technologies more generally). I argue that defining the society we live in as one based on surveillance may actually reinforce political rhetoric in this area. Rather than suggesting that we have reached a certain level of surveillance, it might be more beneficial, in terms of promoting public debate and engagement in this area, to suggest that surveillance technologies have multiple dimensions and purposes, and the uses to which these are put are open to interpretation and change. Arguing that we are already in a surveillance society implies that there is nothing left to resist. I believe there is, and will pursue this argument in the chapters that follow.

The focus for this book is CCTV and video surveillance. However, the wider context of surveillance is also important. As I mentioned in the introduction to this book, I want to move away from thinking about CCTV purely as an inevitable part of the surveillance society that various commentators suggest we have already entered, are entering, or have always been in. In Chapters 3 and 4 of this book I discuss the various uses to which CCTV was put prior to being used as a tool for surveillance, or a tool for politics.

For now though, it seems appropriate to talk about the wider context of surveillance that has developed and evolved over recent years. In this chapter, I therefore provide an introduction to the idea of a surveillance society. I then move on to develop a history of surveillance in Britain.

Surveillance and Society

In twenty-first-century Britain, personal data is collected on the general population every day. Whether in the form of photos and videos taken from smartphones; a supermarket loyalty card; the CCTV cameras on every street corner; the biometric information contained in passports; or the digital cookies tracking your movement online. Personal information is collected in innumerable ways every day by a wide variety of people, technologies, organisations, and institutions. This is not always a passive activity – information is not just taken from people. People also give their personal information away every day in return for products and services, or in order to connect to a range of people on social networking sites. Whether this shows willingness to give away personal data is less clear. Without giving away personal information and identifiable information, that person may be locked out of whatever activity, website, or institution is asking for that information. So, consent is something that is often 'forced' upon people. Of course, you can choose to opt out of using social networking sites but when it comes to opting out of things like providing biometric information for new passports, it is much less straightforward. You can choose not to do it, but you will not be allowed to have a passport. You will therefore not be allowed to travel and that is when consent becomes a complex issue. And of course a personal choice to opt in or out of the use of social media but not travel may not be a decision so easily made by members of Generation Y, whose experience of the world has been digital since birth. Opting out of travel may well be an easier choice for millions of young people in their teens and twenties than opting out of social networking.

The term 'surveillance society' was coined in 1985 by Gary T. Marx, who stated that 'with computer technology, one of the final barriers to social control is crumbling – the inability to retrieve, aggregate, and analyse vast amounts of data. Inefficiency is losing its role as the unplanned protector of liberty' (1985, p.26). This idea that we have entered into a new type of society – one based on surveillance – has been extended by a variety of theorists over the last few years. In his seminal 1994 book *The Electronic Eye: The Rise of the Surveillance Society*, David Lyon argues that this type of society exists due to details of individuals being 'collected, stored, retrieved and processed every day within huge computer databases' (p.3). A few years later, Lyon added to this argument to suggest that as surveillance systems grow, they are becoming less apparent and far more efficient, structured and elusive (2001, p.2).

Theorists in this area of study use examples of information collection, processing and storage to argue that we have entered a surveillance age, or surveillance society. This idea is based on the notion that the changing information flows embedded in the use of new information and communication technologies has led to an increase in surveillance, and the emergence of a new type of society (Gandy 1989; Lyon 2001). Surveillance has become part of the everyday and is embedded in a range of technologies and practices (Murakami Wood and Webster 2009). Others state that it is the rise of computer networks that have meant we have entered into a network age or network society (Castells 1996). These ideas of networks, information and surveillance are also tied together with notions of risk and the idea that we have entered a new era for society – one based on ideas of risk (as I mentioned in the introduction), and strategies concentrating on how to manage this risk.

Surveillance in particular has become a measure used to try to reduce risk. Examples of this can be seen in the widespread use of CCTV. This strategy of managing risk via the increased use of surveillance technologies such as CCTV cameras is coupled with a move towards what has been termed by Norris and Armstrong as a 'stranger society'; a decline of communities and communication (1991, p.29). Young (1999, p.70) argues that this loss of communication leads to 'less direct knowledge of fellow citizens', which leads to 'much less *predictability* of behaviour'. This 'stranger society' is intensified by the development of gated communities, the privatisation of public space, and the widespread use of CCTV cameras (Norris and Armstrong 1999, p.23). Although these developments are implemented in order to manage risk and in the case of CCTV cameras to reduce the public's fear of crime (as I will show in Chapter 5), they may instead have had the opposite effect. Spitzer has suggested that 'the more we enter into relationships to obtain the security commodity, the more insecure we feel; the more we depend on the commodity rather than each other to keep us safe and confident, the less safe and confident we feel' (1987, p.50).

There is a danger with all of the aforementioned theories arguing that we have entered into a new type of society based on technological means (the ability to collect and store greater amounts of information due to improving technological capabilities) that we veer into a technologically deterministic argument – that society is determined by technology. This sort of argument feeds the belief that technologies are unstoppable drivers of change (and that our society becomes based around these), and there is little point in attempting any form of change in relation to the uses of these technologies. I will come back to this issue throughout the remainder of chapters in this book.

Surveillance in Popular Culture

Surveillance has become normalised. We are confronted with surveillance practices and technologies every day and in a variety of ways. As I mentioned earlier, some of these seem to be welcomed, or at least tolerated; others are treated with more caution. We are immersed in a society that has a long history of making reference to surveillance through popular culture. In terms of literature, there are a number of books that make surveillance their main theme, such as: *We*, *1984*, *A Handmaid's Tale*, *The Castle*, *Mistrust*, *The Traveller*, *Discipline and Punish*, *The Birth of the Clinic* – to name but a few. Surveillance has also provided rich material for films, such as: *Gattaca*, *Minority Report*, *THX-1138*, *The Truman Show*, *Code 46*, *Modern Times*, *Metropolis*, *A Scanner Darkly*, *Brazil*, *Caché*, *The Conversation*, *Enemy of the State*, *The Lives of Others*, and *Look*. The references to surveillance and the societies envisaged in these literary texts and films tend to veer towards the dystopian.

To take a closer look at one example, the novel *We* depicts a world in which there are no social classes and only numbers to identify people. In this society, there is 'I' (the benefactor) and 'we' (the collective group of people). Any sense of individuality has disappeared and people therefore have no sense of the self or individual identity. The book is set in the One State – a nation constructed almost entirely of glass, allowing the Bureau of Guardians (political/secret police) to watch the public at all times. The lead character – D-503 – lives in a glass apartment (and all the people live in glass houses). Though far less well known than 1984 (*We* was written in 1920/21 but banned from print in Russia until 1988), it might be argued that *We* is the first piece of dystopian literature, and a book that certainly influenced Orwell. Alongside the themes of bureaucracy, rationalisation, and identity, the idea of a society constantly under surveillance is defined in a strongly negative way.

For Leavis, this is one of the main functions of literature, to give us a 'collective consciousness away from other ideologies in society'. Literature maintains a 'critical function' that allows people to question advances in science and technology (1962, pp.26–9). For Eagleton, this may be too simplistic and he argues that literature cannot necessarily distance itself from the dominant ideology or challenge it, but instead has a particular relationship with it. This relationship means that literature does reflect the ideology of its time but also maintains some distance (allowing the reader to see the ideology from which it derives) (1976, pp.17–18). Hillegas argues that 'quality science fiction' (such as dystopian or anti-utopian literature) makes a significant comment on human society. He argues that it is a vehicle for social criticism as well as satire (1967, p.8). For some, these cultural outlets reflect society's ambiguity towards new technologies (in this case, surveillance technologies). It is sometimes referred to as a form of public resistance (see for example, Staudenmaier 1985, chapter 7).

So, some popular culture maintains a critical stance towards surveillance, aiming to provoke a sense of disquiet or unease about the advancements presented.

However, other areas of popular culture define surveillance in a very different way. For Lyon, TV programmes such as *CSI* provide reassurance about surveillance practices and provide credence to ideas such as 'CCTV works'. Even more than this, programmes such as *Big Brother* 'encourage deliberate disclosure' and 'make full, intimate visibility to watchers a commendable condition' (2007, pp.139–40). I will come back to this idea of popular culture providing a form of justification for the use of surveillance technologies in Chapter 4.

A Brief History of Surveillance in Britain

Pre-modern Surveillance

Although various theorists argue that we have, in recent times, entered a surveillance society, the history of surveillance is a long one. Informal surveillance, or the collection of information on people, can be seen throughout history. For example, Lyon (1994, p.24) cites the Domesday Book in the eleventh century as a primitive form of surveillance. Commissioned in December 1085 by William the Conqueror, the book was completed in August 1086. At the time of completion, it held records for 13,418 settlements, containing information of the amount of land owned, how many people occupied areas of land, and other resources and buildings present on the land. Its purpose was to record the value of land and its assets. Although primarily serving a financial function, the Domesday Book also collected data on England's inhabitants (and for the first time in a comprehensive manner). These sorts of developments in writing and administrative capacity have facilitated changes in governance, and what might be argued to be the birth of a primitive surveillance capability. Lyon argues that these changes are 'highly pertinent to the development of surveillance as a dimension of modernity; printing facilitated the development of modern democratic governance' (1994, p.23).

Surveillance of the general population began to increase during the fifteenth century, with a host of religious organisations collecting information on and keeping records of births, marriages, deaths and baptisms. The practice gathered pace during the sixteenth and seventeenth centuries, alongside accompanying cultural changes as stepping stones toward the birth of modern England, described as an 'extensive Tudor-Stuart practice' (Tricomi 1996, p.18) of informing on religious non-conformism during the early part of the English Reformation (Archer 1993, p.3). Intelligence gathering was carried out by court aristocrats, with a number of institutions and offices uniting or contributing to a 'culture of surveillance' (Archer 2993, p.4)

Although there was an increase in religious surveillance during the sixteenth century, political surveillance became comparatively more important due to the emergence of the nation-state and its 'new needs and a developing capacity to gather and use information' (Marx 2005). Centralised state surveillance emerged with the birth of the nation-state, which made 'extensive use of undercover techniques to protect their political, military and economic interests' (Fijnaut and Marx 1995, p.8). Surveillance also became an exclusionary force during the sixteenth century, with the government collecting information on those it felt threatened by:

> Increasingly intransigent and expensive political (religious) and economic problems with the low countries and Spain, including tense military stand-offs in the 1580s, coupled with a fourfold increase in the national population between 1500 and 1600, resulted in rising prices of essentials, food scarcities, popular dissention and riots, an increase in the proportion of the London poor greater than that of the city's population, collective Xenophobia about aliens, and the 'obsessional' surveillance that was the response of the government to perceived challenges to its authority and to fears about alien infiltration and corruption of its national life. (Habib 2008, p.118)

During the seventeenth century, public health information relating specifically to morbidity and mortality started to be collected for the first time, due to a 'fear of plague epidemics' (May 1991). The Office of the Registrar General was established during the eighteenth century, collecting data on births and deaths in England, and continuing the trend of surveillance of diseases. It was not until the nineteenth century, however, that this surveillance of diseases and mortality had 'evolved as a means of collection and interpretation of data related to environmental and health monitoring processes for the definition of appropriate action; for prevention and healthcare' (Goodwin Gerberich et al. 1991, p.163).

More widespread and regular information collection about the population, rather than a focus on births and deaths began with the first Census in 1801 (which was then repeated in 1811, 1821 and 1831). However, the information collected was not particularly detailed. The data might have included the name of the head of the household and their occupation, and possibly the number of males and females in the household. The first detailed national Census was taken in 1841, which gathered more detailed information, including: name, age, occupation and place of birth.

Modern Surveillance

For many theorists, modern surveillance is linked to industrialisation, capitalism, and the growth in administration of the nation-state and modernity. For example, Lyon (1994, p.24) argues that 'systematic surveillance … came with the growth of military organization, industrial towns and cities, government administration, and the capitalistic business enterprise'. I will come back to policing in more detail later in Chapter 3 but briefly touching upon it now, surveillance of the public began to be a part of the everyday life of a policeman and the public:

> The new police received an omnibus mandate: to detect and prevent crime, maintain surveillance of the daily life of their areas and report on political

> opinions and movements, union activities and even recreational life ... Upon their introduction they made themselves obnoxious by imposing more vigilant surveillance of public houses, cracking down on foot-racing, wakes and fairs and introducing the novel and hated 'move-on' system. (Storch 1980, p.34)

The police operated on foot during this time, known as the 'beat system'. This system 'presented an intrusion into working class neighbourhoods not previously kept under routine surveillance. Their interference into the lives of the poor was unsolicited and deeply resented – most strongly ... in the "move on" system as an attack on the traditionally sanctioned freedom of assembly on the streets' (Bailey 1981, p.74).

Surveillance of the population at this time (late nineteenth century) therefore represented not only a monitoring of citizens but also an exclusionary tactic. A growth in urbanisation meant that those who did not play an active role in civil society were put under increasing surveillance. A number of criminal registries were set up during this time, with the introduction and inclusion of photography, body measurements and fingerprinting from 1871 onwards (De Leeuw and Bergstra 2007, p.9). Dandeker (1990) characterises developments during the nineteenth century as one of 'great transformation'. Cohen (1985, p.13) agrees and argues that during this time 'master patterns and strategies for controlling deviance in Western societies' were established. Thus there was a move during the latter half of the nineteenth century towards a more centralised and modern form of surveillance (Cohen 1985).

Centralised state run surveillance really began in earnest at the start of the twentieth century. In England the 'combination of warfare and welfare' created the basis of the surveillance we see nowadays (Lyon 2003, pp.23–4). Improvements in transport also increased the need for national forms of identification, such as the driving licence, which was introduced in 1903 (De Leeuw and Bergstra 2007, p.9). 1915 saw the introduction of the passport, which included details of the individual such as name and date of birth, a photograph and signature. Surveillance had by this point become a way to prove citizenship, rather than as a method of exclusion for certain elements of society, which it had mainly been prior to this time. In this sense surveillance had become an inclusionary, as well as an exclusionary phenomenon.[1]

It did still remain an exclusionary method, however. One of the earliest examples of state surveillance was the photographing of the Suffragettes. It is likely that this is the earliest example of an activist, or what some commentators have deemed terrorist, organisation being subjected to covert surveillance

---

1 When I talk about exclusionary methods of surveillance I am referring to the introduction of routine surveillance of working class areas by on the beat policemen, which began during the early nineteenth century. Alongside this routine surveillance, Storch (1980) describes the introduction of the 'move on' system, which hindered the traditional freedom of assembly on the streets.

photography. The photos were a mix of studio portraits, press pictures taken at public demonstrations, police photographs taken at the time of arrests and those taken in the exercise yards at Holloway Prison. The photographs taken in the exercise yards were taken covertly. Forced photographs in other situations were also taken (Hamilton and Hargreaves 2001, p.55).

The first British Identity Card was introduced in 1915 and abandoned in 1919. The second was introduced in 1939, with the system in existence until 1952. The first identity card was used for the purpose of national registration, in order to determine the number of males in the population available for conscription (Agar 2005). Agar points out that this was not the reason why the card was introduced, yet it did become the reason for its use. The original intention had been to join up numerous personal registers of information on individuals held by government agencies. The second iteration of the card met three purposes, specified as being 'for the duration of the present emergency: co-ordinating national service, national security and the administration of rationing' (Agar 2005). Once the war ended there was significant public and political opposition to the cards, which were withdrawn in 1952.

During the 1960s there were a number of changes in police use of CCTV, which I cover in more detail in Chapter 3. In terms of more general changes, plans for a Police National Computer (PNC) were announced in 1969 (Manwaring-White 1983, p.55). The PNC was launched in 1974, providing each of the 47 police forces in England and Wales with one terminal from which to access the central computer (ibid.) The PNC is now 'linked to more than 30,000 terminals across the country', including databases run by the DVLA, the National Automated Fingerprint Identification System, and the Violent Offender and Sex Offender Register (Norris 2007, p.144).

ANPR cameras began to be developed during the 1970s. The Police Scientific Development Branch designed the first prototype in 1976. By 1979 several prototypes were being tested on the A1 and in the Dartford Tunnel. During the 1980s, ANPR was used in support of counter-terrorism initiatives (used covertly during this time). However, the technology proved to be very expensive (Sheldon and Wright 2010, p.126). Improvements in the technology, and a decrease in cost, meant that by 1999 fifteen police forces in Britain were using ANPR systems (ACPO 2004, p.2).

The Labour Party, with Tony Blair as leader, won the 1997 general election, and subsequently winning in 2001 and 2005. The leadership of the Labour Party changed hands in 2007, passing to Gordon Brown, former Chancellor of the Exchequer. Under Tony Blair there were a number of notable changes in human rights legislation, with the Human Rights Act 1998 brought into force in 2000, promoting further the rights and freedoms granted to citizens under the European Convention on Human Rights. However, during this period there were also developments in the area of surveillance of individuals and the collection of personal information, with the Identity Cards Act 2006 planning the introduction of a new National Identity Card, linked to a National Identity Register database (a scheme which has now been abandoned).

Since the late 1990s and into the early twenty-first century, there has been a huge increase in the number of government-run databases in existence, as well as an increase in the amount of information stored in them. These have ranged from databases to determine eligibility for services, to those designed for welfare reasons. The plan for the previously mentioned National Identity Register (linked to the National Identity Card proposal) was to store a vast amount of personal information, with the Labour government stating that it would hold the data (including biometric information) of all British citizens on the system. Of this development, the Information Commissioner at the time, Richard Thomas, stated: 'My anxiety is that we don't walk into a surveillance society' (BBC News, 16 August 2004). Plans for the National Identity Register and identity card scheme were finally scrapped in February 2011 under the new coalition government, with the Deputy Prime Minister, Nick Clegg, stating: 'The ID cards scheme was a direct assault on our liberty, something too precious to be tossed aside, and something which this government is determined to restore. The government is committed to rolling back as much state interference as humanly possible, and the destruction of the register is only the beginning' (Home Office Press Release, 10 February 2011).

The UK National Criminal Intelligence DNA Database was set up in 1995 and now holds profiles of over 5 million people (Home Office 2013). DNA samples are taken from crime scenes and from individuals in police custody and their profiles held on the database. The profiles of 5.2 per cent of the British population are on the database, making it the largest of any country in the world. According to the Home Office, over 300,000 crimes have been detected with the 'help of the database' since 1998 (Home Office 2013). Of course there are major issues tied in with developing a database of this kind. A range of ethical issues have been brought up over the last few years, ranging from those focusing on privacy and civil liberties, to those focused on the robustness of DNA evidence can provide. Often seen as a kind of infallible 'truth' when submitted as evidence in court, there is a limit to what DNA evidence can prove.

The database, ContactPoint, was set up in 2008 to store information on every child in England under the age of 18. This database was created under the Children Act 2004, an amendment to the 1989 Act, largely in response to the Victoria Climbié inquiry.[2] The purpose of the database was to allow the sharing of information between children's service providers. The scheme was disbanded under the new coalition government in 2010. The planned NHS National Programme, designed to store all NHS patient records in one central electronic database, was also discontinued in 2010 under the new coalition government.

---

2 Victoria Climbié was an 8-year-old girl, tortured and murdered by her guardians. Prior to her death a range of people from social services, the NHS, local churches, and the NSPCC, had contact with her.

Political campaigners have faced increasing levels of surveillance in the last few years, with their details stored on the database Crimint, used by the police to catalogue criminal intelligence, which includes: criminals, suspected criminals and protestors (*The Guardian*, March 2009). The information collected, which includes names, political meetings and/or demonstrations attended, is kept for up to seven years. Furthermore, photographs and video surveillance footage of campaigners (often taken covertly) are also stored on the database. Political protest is not necessarily a criminal offence; however, political campaigners are subjected to pre-emptive surveillance. A Freedom of Information request from January 2012 shows that there are approximately 14,000,000 records on the Crimint system, most of which contain text, although there is the possibility to attach an image file as well. The FOI request details the use of Crimint by 40,000 users in the Metropolitan Police Service, and states that it is used for the purposes of supporting 'a full range of policing activities from neighbourhood policing to investigations into serious and organised crime' (Metropolitan Police 2012).

The use of ANPR has also increased substantially since the late 1990s. Project Spectrum was launched in June 2002, making £4.65 million available to equip every police force in England and Wales with a mobile ANPR unit (and associated software necessary to match and store ANPR data) (Sheldon and Wright 2010, p.127). In 2004, the Home Office announced a further £15 million to expand the use of ANPR by the police, as well as for the creation of a national ANPR data centre in Hendon (ibid.). In 2005, the Association of Chief Police Officers (ACPO) stated that: 'the intention is to create a comprehensive ANPR Camera and Reader infrastructure across the country' (ACPO 2005, p.17).

Surveillance in airports and other transport hubs has also increased in the last few years. Airport security in particular has turned to electronic security measures rather than relying on human intervention. E-passport gates have been installed, and are in use, at all the major UK airports (15 in total). These automated border controls utilise facial recognition technology to match a passenger's face with the photograph stored on the biometric chip in their passport (although a human immigration officer is usually present at these e-passport gates). Other details stored on a passport are also checked against the UK Border Agency systems. Prior to the installation of e-passport gates, a number of iris recognition immigration systems (IRIS) were installed at UK airports (the first IRIS gates were installed in 2004). These systems worked by matching a passenger's iris to a database. However, the system proved to be ineffective and inefficient, with passengers taking longer than the estimated 12 seconds to make their way through the gates. Interestingly, in this context biometrics were seemingly installed for the purpose of efficiency and speed, rather than security purposes (for a history of biometrics in terms of speed and security see Bright 2011). The IRIS scheme has now been stopped in the UK and the government are in the process of decommissioning the system. Since 30 September 2012 the gates have only been used for registered passengers in Heathrow Terminals 3 and 5 (Home Office 2013).

Body scanners ('backscatter' scanners) began to be trialled in 2009 at Manchester Airport. Questions began to be raised in 2011 by the European Commission regarding how safe these scanners were in terms of levels of radiation emitted. Backscatter scanners were eventually ruled to be unsafe and disbanded for 'health and safety reasons'. Alongside these health and safety concerns, 'privacy issues' were also cited as a reason for the introduction of a new type of body scanner, which does not involve a human checking the image, rather the scanner shows a cartoon-like image. However, if there is cause for suspicion, a body search will be conducted.

Biometrics, and the use of bodily characteristics as identifiers, is not a new phenomenon (for example, Jain (2004) describes body measurements being used as identifiers during the nineteenth century). However, recent advances in digitisation processes (i.e. the capturing and processing of personal data) have meant that biometric technologies now have the potential to be automated (and to become, in theory at least) more efficient and cost effective. Over recent years, the UK Home Office have described a number of ways that biometrics could be used, including:

- 'the fight against illegal working' by preventing employers employing staff who do not have the right to work in the UK and also including a biometric work visa for non-permanent residents;
- 'immigration abuse' by making the UK less attractive for asylum seekers as, unlike some other EU states which have ID cards, many come to the UK and disappear;
- 'the use of false and multiple identities' by terrorists and criminals;
- ensuring 'free public services are only used by those entitled to them – preventing abuse such as "health tourism"';
- helping to 'protect people from identity theft' whereby victims have their identities stolen by others who may use the identity for financial or some other gain (according to the Home Office this could 'touch' one in four UK adults and it takes victims an average of 300 hours to 'put their records straight') – and ID cards could speed up this identity reconciliation.

Proponents of biometric identification systems argue that the inclusion of biometric information makes it very difficult to falsify characteristics, which in turn makes a secure and robust system. However, the governance of these systems is complicated for a number of reasons. There is a lack of knowledge about the technology; it is complex and requires a certain level of knowledge and expertise to understand it (Bright 2011). There are a variety of competing versions of biometrics at present (no one definition of the technology exists) and there are vastly differing claims as to its effectiveness, accuracy, how intrusive it is, and even what it should be used for. Research has shown that the definition and uses of biometrics have shifted over recent years. Pre-9/11 they tended to be used for reasons of efficiency and speed (at least in theory), shifting to uses of security post-9/11. They are now shifting again to become a technology to be used for access or entitlement (ibid.). How they will eventually be used in the UK remains to be seen.

This emphasis on collection of personal information and storage on databases over recent years has coincided with a rise in the importance of targeting crime at the local level for the police. In 1998, the Crime and Disorder Act was created, which Newburn (2008, p.95) describes as placing a statutory duty on chief police officers and local authorities, to work in cooperation with other authorities (such as health authorities and probation committees) to 'formulate and implement a "strategy for the reduction of crime and disorder in the area, including undertaking and publishing an 'audit' of levels and patterns of crime locally"'. He adds to this by stating; 'the language of the late 1990s was dominated by talk of "partnership", of multi-agency working and of joint responsibilities, and it was in the area of community safety that this was perhaps most visibly seen' (ibid.). A number of Community Safety Partnerships (CSPs) were set up by the Home Office in 1998, designed to focus on local issues such as anti-social behaviour, drug or alcohol misuse and reoffending. They are made up of representatives from: the police, local authorities, fire and rescue authorities, the probation service, and the health service. The Home Office (2013) states that CSPs 'annually assess local crime priorities and consult partners and the local community' regarding how to combat the local crime priority issues. This focus on partnership and communities has continued in the early part of the twenty-first century. In their 2004 White paper 'Building Communities, Beating Crime', the Home Office emphasised good police-public relations ('a new relationship between the police and the public – trust and confidence') as an essential part of policing practice.

The Metropolitan Police Service Communities Together Strategic Engagement Team (CTSET) was formed in 2005, in response to the terrorist attacks on the London Underground on 7 July of the same year. The main role of the team is described as being 'responsible for engagement and consultation with the Met's key strategic partners, stakeholders and networks, as well as London's diverse communities, within the context of counter terrorism and security' (Mayor's Office for Policing and Crime 2013).

Anti-terrorism work has become a more major role for the police during the early part of the twenty-first century. With regard to police training for terrorist attacks, Edwards (2005, p.292) states that it may only be 'extremely expensive window-dressing to persuade the public that police are doing something to counter the fear of terrorist attacks'. The Office for Security and Counter-terrorism sits within the Home Office and works in four main areas; outlined in the Government's counter-terrorism strategy, CONTEST. They state that they work to: 'stop terrorist attacks; stop people becoming terrorists or supporting terrorism; strengthen our protection against a terrorist attack; mitigate the impact of a terrorist attack' (Home Office 2013).

Over recent years, a range of anti-terrorism legislation has been passed, alongside a range of surveillance activities, justified through the aim of combating or fighting terrorism. In October 2010, the Government published its new National

Security Strategy, identifying terrorism as one of the top four threats faced in the UK today. A range of legislation has arisen from this anti-terrorism initiative, including: the Terrorism Act 2000, the Terrorism Prevention and Investigations Measures Act 2011, and a draft Communications Bill (2012).

Various voices have spoken out against this increase in anti-terrorism legislation, as well as the intermittent use of 'exceptional powers', when civil liberties are restricted under the banner of 'preventing terrorism'. Civil liberties groups such as Human Rights Watch (2011) have argued that the use of exceptional powers needs to be curtailed and brought in line with basic human rights standards. Others have taken this argument one step further, however, and state that the preceding argument rests on the belief that there is a current and real threat. For these commentators the threat of terrorism is yet to be defined and proven:

> Governments are able to do so in no small part because of the semantic fog that surrounds the core concepts of national security, threat and terrorism by which exceptional powers are usually evoked. Terrorism, for instance, is a concept that resists consistent definition. Commonly understood by governments as the use or threat of use of serious violence to advance a cause, the term elides legitimate resistance to occupation and oppression with 'senseless destruction'. Furthermore, by relegating all terrorists to the criminal sphere, the term delegitimises any political content that acts regarded by authorities as terrorist may have. This helps to obscure from the public the reasons why people resort to such acts. It also enables the police character of the proper response to be presumed. (Anderson 2012)

In 2012, a draft Communications Data Bill (known informally as the 'snooper's charter') was proposed by the Home Secretary, Theresa May, including plans for mobile phone providers and internet service providers to collect and store information on users' internet activity, email communications, mobile phone messages, and phone calls. The ideas included within this proposed bill are not new. Under New Labour there were plans for similar activities under the Interception Modernisation Programme. Deputy Prime Minister, Nick Clegg, withdrew his support for the draft bill in April 2013. However, following the Woolwich attack in May 2013 there has been increasing political pressure on Clegg to revive the bill.[3]

Conclusion

As we have seen in this chapter, the history of surveillance in Britain is linked to various developments: welfare services, conscription, travel and citizenship. It has now become more and more difficult not to have our images and movements

3 On 22 May 2013 a British soldier, Drummer (Private) Lee Rigby was attacked and killed by two men in Woolwich, London. The incident has been treated as a terrorist attack.

captured many times a day, our shopping habits tracked, our daily journey to work captured and stored on a database. However, the view that we are now already living in a surveillance society implies that there is nothing left to resist – that this is the society that we live in and there is not much that we can change. In 2008, Tony McNulty (the then Minister for Security, Counter-terrorism, Crime and Policing) stated that: '[Surveillance] is today's normality. CCTV, DNA database and a whole range of these other elements are not there as a response to exceptional threats and exceptional circumstances … I think that is routine in the 21st century' (House of Lords Select Committee 2009). I agree that surveillance has become normalised. However, to state that it has simply become 'routine' is to leave questions unanswered. It implies a form of acceptance – it is simply routine, mundane, and the way things are. Why have surveillance technologies become so widespread and utilised? Why has there not been more public pressure on the government to ensure that these systems and technologies are used proportionally and for specific purposes? How have ideas of security, terror and crime been constructed as to allow these systems and technologies to become a part of the everyday? By looking at CCTV as an example, this book will start to question assumptions and ideas around surveillance.

Before looking more specifically at the history of CCTV I have provided an outline of how surveillance technologies and practices have developed in Britain in recent years. As has been shown, there has been an increase in technological means and aids to surveillance, however not all government run information technology and communication technologies should be thought of as purely surveillance systems. They are also developed and utilised for other purposes. As well as questioning the assumption that we live in a surveillance society in Britain, I also want to question the assumption that CCTV is unambiguously a surveillance technology. CCTV has been argued to be 'explicitly about surveillance' (Webster 2012). In some ways it becomes synonymous with the surveillance society (at least in popular rhetoric). When CCTV is situated within this type of argument, it is easier to lean towards a thesis of inevitability – that it was destined to become a surveillance technology and this is the only purpose for which it was developed.

In the following chapters I develop a history of CCTV in the context of social change in Britain, changes in politics, policing, criminal justice, science policy, and public engagement. If we are to answer the question of why Britain has become so camera-surveilled, we need to understand the social, economic, and political factors and influences that have helped to construct it into a technology used for surveillance in public space.

Chapter 2
Post-war Britain and the Public

Introduction

> The Britain that is going to be forged in the white heat of this revolution will be no place for restrictive practices or out dated methods on either side of industry. (Wilson 1963)

In the post-war era, significant cultural shifts and social changes have occurred in Britain. Alongside a major growth in the population, there have also been important changes in the welfare system, immigration, social housing, and fluctuations in the economy and a relative rise in poverty. There have also been important changes in science and technology. Scientists in post-war Britain enjoyed a period of relative autonomy in respect to their research, and a growth in levels of funding. The idea of science and technology as synonymous with 'progress' also gained pace during this time. This move towards the idea of science and technology as progress has led to the idea of a techno-fix for certain social and environmental problems.

As I have mentioned in the introduction to this book, I believe that the history of CCTV in Britain needs to be looked at in the wider context of social change, the economic and political context in which it developed, policing and criminal justice, and the regulatory environment. In this chapter, I develop a social history of Britain since the end of the Second World War. The chapters that follow look at the history of criminal justice and policing (including the use of CCTV by the police) and the more general history of CCTV in Britain in the same time period. My focus here is on developments in politics, the economic context, social change, and changes in science and technology, as these areas all have implications for the history of video surveillance in Britain.

Post-war Britain

The End of the War

Winston Churchill had been the British Prime Minister for the majority of the Second World War, leading a coalition government (he was a Conservative). This wartime government came to an end on 23 May 1945 (after the end of the war on 7 May 1945). Clement Attlee, and a Labour government succeeded Churchill. Sked and Cook (1993, p.18) described the change from a Conservative to a Labour government as surprising at the time, however explain the change in terms of people assuming after the war that they would 'share in common rewards … in better housing and social services' and that 'they knew these benefits were more likely to be provided by Labour than the Tories'.

In the aftermath of the Second World War, Britain has been described by Hollowell (2003, p.61) as being 'economically exhausted', facing continuing rationing and near bankruptcy. Marwick (1996, p.7) described Britain as a country seeking a 'new future involving state planning and a comprehensive welfare state'. In terms of greater state involvement, the coal, gas, transport, and electricity industries were nationalised under Attlee's government (Glynn and Booth 1996, p.268). The National Health Service was established on 5 July 1948, providing universal and free healthcare (Mullard 1995, p.183). Production levels rose during this time, reaching its highest ever level by 1950. Exports also increased – by 77 per cent between 1945 and 1950 – and reached their pre-war level by 1946 (Childs 1995, p.89). Despite this rise in production levels, most basic foods remained rationed and there were demands from the middle-class, coming to the end of an almost full decade of employment, for higher standards of living. This demand for improved standards of living, coupled with a feeling that levels of taxation for the middle-class were too high, meant that support for Labour decreased (ibid.). Labour did, however, win the 1950 election, with Attlee remaining as Prime Minister. However, their majority had significantly lessened, with only a residual five-seat majority. The following year, in October 1951, Winston Churchill became Prime Minister once again under a Conservative government. The Conservatives remained in power from this time until October 1964.

Under Churchill (October 1951–April 1955) domestic policy took a back seat to foreign policy, due to Britain's involvement in the Korean War (MacDonald 1990). There was no return to the pre-war laissez-faire and the welfare state remained in place. Alongside the enduring welfare system, the majority of industry remained nationalised. Despite the Conservative government facing a 'serious balance of trade situation', the early 1950s were a time of increasing affluence (with full employment until the 1970s) (ibid.). There was also a sense of increasing freedom, which had been an election promise from the Conservatives prior to the 1951 election when they stated that they would 'set the people free' (Francis 1996). Broadcasting was deregulated in 1954, and following this, commercial television was introduced in 1955.

The period from 1953 to 1973 has been called the 'golden age of prosperity', a time of massive economic expansion across Europe (although for Britain the economic expansion occurred primarily after the war and during the early 1950s). At this time there was a rise in production and consumption, an increase in jobs, and rising standards of living. Wakeman (2003, p.45) argues that 'a consumer revolution swept through Europe completely transforming everyday culture with a cascade of commodities from washing machines to telephones and televisions'. The average wage rose by 34 per cent between 1955 and 1960, whilst the cost of consumer technologies fell. However, for Britain this economic prosperity had started to wane by the mid-1950s (Holden 2005, p.35).

The 1960s

Against this background of economic and cultural shifts, there were also a number of changes in social relations. Sir Philip Knights suggests that:

> Not only did the police get into cars, but so did the public. It was only one of the new problems of a society, which changed enormously in the '60s. We used to talk to people over the garden fence, now the fence is twenty storeys up in a high-rise building. It wasn't only the police withdrawing, but society developing in an insular way. (Sir Philip Knight quoted in Manwaring-White 1983, p.23)

This withdrawal of society occurred alongside a rise in relative poverty. Although, as described earlier, there was a rise in wages and a fall in the price of consumer goods, the gap between those gaining more affluence and those with little widened and became more pronounced throughout the 1970s. Housing estates, which had been built during the 1950s and 1960s to provide a solution to a lack of housing after the Second World War, had become, Marwick (1996, p.232) argues:

> Heartless public housing estates … now stuck on a descending spiral into Hades as, a natural target for frustration and vandalism, they increasingly became dumping grounds for problem families.

Hanley (2007) describes council housing in Britain as having become 'housing for the working class (and the non-working class)' and that 'a wall exists unbroken throughout every estate in the land'. This divide, which started to occur in the 1950s and 1960s in Britain with the building of high-rise and sprawling council estates, meant that certain sections of society were becoming more isolated and detached. By the mid-1960s a quarter of all council housing approvals were made up of high-rise blocks over nine storeys (although this fluctuated by area and was considerably more in some districts). However, this did change during the 1970s with a number of voluntary and charitable agencies voicing concerns over the social problems emanating from, for example, bringing children up in high-rise buildings (Ravetz 2001, p.104). Ravetz argues that it was during this time that 'even many architects "stood appalled" at what their profession had apparently unleashed' (ibid.).

The 1970s

During the 1970s, numbers of trade union memberships grew substantially. This continued until 1979. During the late 1960s and early 1970s, alongside a rise in inflation, there was also a return to major mining disputes, culminating in a period of 'higher than usual levels of strikes', in particular in the coal-mining, docks, shipbuilding, car manufacturing, and iron and steel industries (Wrigley 1997, pp.26–9).

By the mid-1970s the British economy was showing signs of distress. Sked (2003, p.48) argues that the economic failures of the 1960s and early 1970s including: 'growing unemployment, rising inflation, balance of payment crises, the 1967 devaluation and the 1976 IMF crisis', meant that 'people began reassessing the assumptions behind the consensus'. Furthermore, he suggests that the voting public viewed the trade unions as being responsible for Britain's economic decline. The subsequent drop in support for trade unions allowed Thatcher to openly campaign against them in the general election of 1979, eventually winning a mandate to reform them.

The Conservatives, with Margaret Thatcher at the helm, won the 1979 general election. During the previous years, with Labour in power and James Callaghan as Prime Minister (1976–1979), relations between the government and trade unions had worsened and eventually broken down during the Winter of Discontent (1978–1979). Ironside and Seifert (2003, p.47) argue that the Winter of Discontent was caused by Labour's 'shift to the monetarist right ... and its turn away from social democracy', which in turn paved the way for the election victory of the Conservatives under Thatcher.

The 1970s and into the 1980s

The Thatcher government stayed in power until 1990 (winning a further two elections in 1983 and 1987). During her three terms, Thatcher introduced increases in Value Added Tax (VAT), cuts in direct taxation, and decreases in public spending (Black 2004). There was also a move towards deindustrialisation and privatisation of industry, reducing the activities undertaken by the state. Deregulation was supposed to increase competition, and it was for this reason that private businesses were encouraged. Concurrently, there was a move to gradually reduce eligibility for state benefits, in an attempt to encourage people to return to work. However, at this time there was also a rise in unemployment, particularly in the manufacturing industries (Childs 1995).

During the late 1970s and into the 1980s social divisions deepened in Britain with regard to wealth, race, and class. Throughout the 1980s there were also a number of social and political disturbances (Marwick 1996). For example, in 1981 there were a number of outbreaks of rioting in London, Liverpool, and Manchester. For some these riots reflected local issues between residents and the police. For others the riots told of a deeper sense of alienation from society. A few years later, the government faced a year-long strike by the mining industry (I return to this in Chapter 3). Throughout the 1980s there was also a continuing threat of violence from the IRA (the armed paramilitary campaign had started in 1969 and continued through the 1970s). During the 1984 Conservative Party annual conference an IRA bomb was planted in the venue. Five people, including one Member of Parliament (MP), were killed (this incident has particular relevance for the history of CCTV, and I return to this issue in greater detail in Chapter 4).

By 1987, the state sector had been cut by a third. Fourteen large companies had also been privatised (mostly public utilities, and telecommunications), amassing £11 billion for the Treasury. Continuing this path of privatisation, the Conservatives also brought in legislation designating a 'right to buy' for council tenants wishing to buy their homes, allowing them to do so at a discounted price. This increased the number of homeowners substantially, yet, at the same time, meant that fewer council houses were available for others. The number of council houses fell, increasing the number of homeless people (Zander 1990).

The 1990s and Onwards

The Conservatives retained power in the general election of 1992, with John Major as Prime Minister (having won the leadership election in 1990). They remained in power until 1997. By the time Major stepped in as Prime Minister, Britain had gone into recession. Unemployment at this time was also rising, after a number of years of improvement. This situation began to change in the latter half of the 1990s, with employment levels rising once again, and an increase in weekly earnings (Llg and Zaugen 2000).

During the 1990s there were also major advances in Information Technology (IT), computerisation and the storage and collection of information on a mass scale. These changes had been occurring since the end of the Second World War in terms of analogue technology, which enabled the storage of data for the first time. This in turn was superseded by the advent of digital technology, which allowed for greater volumes of data to be stored at lower cost. Technologies such as the microchip and GPS have allowed for tracking of individuals, as well as the advancement of other surveillance technologies. In terms of computer technology, advances have been threefold: Firstly, the amount of information stored has increased, secondly, the speed at which this information can be transferred has increased and thirdly, improvements in search facilities have occurred, allowing specific information stored to be found more easily.

The early part of the twenty-first century has also seen further advances in computerisation, digital technologies, and the Internet, leading to changes in the workplace, consumption, and everyday interaction with others. The Internet has also allowed (at least in theory) an increase in public participation in decisions on science and technology, in terms of online consultations, citizen participation exercises, and other forms of E-democracy. There has also been an increase in government run databases as I mentioned in the previous chapter, allowing increasing amounts of information on the population to be stored, collected and processed.

Science and Technology Policy

The post-war period in Britain has been described as a 'golden age' for science and technology (Keenan and Flanagan 1998). A number of research programmes were set up in the fields of aviation, atomic energy and electronics, and government support for these areas, as well as others such as space research, increased well into the 1950s. Resources for science and technology (although the emphasis was mainly on science, which it was assumed would lead directly to growth in innovation) have been described as growing at an 'unprecedented rate' during this time (Keenan and Flanagan 1998).

During this post-war period, science was defined as a 'motor of progress'. In 1963, Harold Wilson gave his famous 'White Heat' speech in which he stated that: 'The Britain that is going to be forged in the white heat of this revolution will be no place for restrictive practices or out dated methods on either side of industry'. With this speech, he cemented a turn towards technological innovation, and more resoundingly a technocratic ideology (which it could be argued still resides in government policy on science, technology and innovation to date). The steadily increasing science budget also gave rise to 'Big Science' (to simplify: large scale projects with large teams of scientists and researchers).

During the late 1960s and into the 1970s, social concerns over the effects and impacts of science began to surface. Publications such as Rachel Carson's *Silent Spring* brought the environmental costs of scientific research to the publics' attention (this book is credited with starting the environmental movement). By the mid-1970s economic pressures in Britain meant that funding for science and technology began to decrease, with cuts to spending and an increasing pressure for greater social responsibility in terms of the potential impacts of scientific research. During the 1980s, industrial innovation (rather than university-based research) became the focus for research policy. Policies based on selectivity and prioritisation took precedence, and science funding entered a 'steady state'.

Throughout the 1990s and into the early twenty-first century, there has been an increasing emphasis on 'wealth creation' and 'quality of life', as well as a growing emphasis on what came to be known as the 'Public Understanding of Science' (I return to this in the next section). The new coalition government has committed to reducing the size of the national deficit through cuts in public expenditure. The scientific research budget has been maintained and 'ring-fenced'. Often cited politically as a positive, when taking inflation into account, this means a cut in real terms.

The idea that science leads to technology, which in turn leads to economic growth and wealth creation, is an idea that has been disputed widely in the academic literature. This 'linear model' of innovation is worrying, and it is unfortunately alive and well in a number of recent governmental White Papers. This model depicts changes in science as leading to changes in technology, and in turn changes to the economy and society. The 1993 Annual Review of Government funded R&D, stated:

> Science can be defined as a systematic study of the nature and behaviour of the material and physical universe. Technology is the practical application of this knowledge.

There are numerous other examples of this sort of rationale for the promotion of science and technology, including:

> The Government is committed to creating a society that is excited about science and values its importance to our social and economic well-being; feels confident in its use. A Vision for Science and Society. (Department of Innovation, Universities and Skills 2008)

> Growth is the Coalition's highest priority for 2012 … A lot of it will come from our established, high performing sectors which continue to innovate and improve productivity. We can discern already some of the scientific discoveries and technologies that will shape our future. (Rt. Hon David Willetts MP, Minister of State for Universities and Science 2012)

Of course there are benefits for scientists to this model. It means that they are autonomous; they choose the problems and then set about solving them, and have the resources to do so. Scientists are also seen as having some sort of superior expertise in terms of problem solving, and solutions to social problems. Science, leads to technology, leads to the solution. Science (and in turn technology) is seen as robust and infallible. The end result often becomes a techno-fix, and this is why this is interesting for the history of CCTV. In Chapter 5 I discuss the idea of a techno-fix in relation to empirical work conducted on the Digital Bridge project. As I will show in this chapter – rather than focusing on the possibility of tackling social problems (or socio-economic issues) with a range of approaches – the tendency becomes one of reaching for a technological solution. In Chapters 3 and 4 it will also become apparent that when thinking through the history of CCTV in terms of policing and politics, CCTV can be thought of in terms of a techno-fix (and often situated outside of any sort of socio-economic context). In the case of CCTV – crime happens, and rather than seeking to address the underlying issues that contribute to certain individuals or groups in society committing crimes in the first place, the answer instead becomes one of reaching for the technological solution or plaster.

Science, Technology and the Public

Under the response of a techno-fix, 'the public' do not play a part in the process of either consulting on the issue, or becoming part of the solution. Historically speaking the public are defined as passive, uninformed and unengaged recipients, who need to appreciate the worth and value of science and technology. Rather than a two-way dialogue about possible benefits and disadvantages, science and technology stand at the top of the hierarchy (with politicians and policy-makers in the middle), and the public sit at the bottom of the pile (apparently reaping the rewards of science and technology). The 1985 Bodmer Report is credited with starting the Public Understanding of Science movement. The general idea behind this report was that if the public understood science more, then they would learn to love, appreciate and support it. In turn, funding for science would increase. Although the academic literature, especially in the area of science communication, has widely critiqued this idea, it still seems to reside in a lot of the policy papers on science, technology and innovation in recent years. It is also prevalent in discourse on CCTV. I come back to this in Chapters 4, 5 and 6.

The main problem with this model is that it eradicates any discussion and debate on risk. It also stops the public being able to ask what science and technology is for (in this case CCTV), who is going to benefit from it, and for what purposes is it going to be used (Wilsdon and Willis 2004). In terms of the history of CCTV, the technology began to be utilised in public space during the 1960s. This utilisation began to accelerate during the 1980s. During this period there was little public engagement on science and technology and therefore little forum for the public to become involved in decisions over how and where to use the technology (and even whether it was the right approach in the first place).

In terms of involving the public with science and technology, scientists and policy makers first sought to educate, and then moved towards dialogue and participation. Wilsdon argues that this is now changing once again to a new mode of engagement – one that is 'upstream'. During the second phase of engagement, ideas of science and society became important and a House of Lords report argued that there was a 'new mood for dialogue'. The newer shift towards an upstream engagement has involved a rise in consultation papers, citizen juries, focus groups and stakeholder dialogues (Wilsdon and Willis 2004).

However, it is fair to suggest that the vast majority of engagement has taken place on environmental issues and has been somewhat different in areas of security and surveillance. I will come back to this later in the book, and will argue in Chapter 5 that the public are severely under-represented in discussions on CCTV. They are still depicted and presumed to be passive recipients of a technological solution. Although the 1990s and early twenty-first century has seen a shift in terms of relationships between science, technology and society – with a change in, or end to, traditional top-down information policies, and greater public participation in policy processes on science and technology – this change has been minimal in the area of surveillance and security (and particularly video surveillance). Although the rise of CCTV began at a time of little public engagement in science and technology, that is no longer the case. Yet consultation and engagement on CCTV has not evolved at the same rate as in other areas (such as environmental issues).

Conclusion

In this chapter we have seen that Britain has undergone a series of changes since the end of the Second World War. A period of relative economic stability in the post-war era, and a rise in consumption and consumer activity, declined during the late 1960s and early 1970s. Against a backdrop of a relative rise in poverty, social relations changed for the police and the public. Society started to develop in a more insular way resulting from changes in urban housing and a number of high-rise estates being built. These high-rise solutions led to a number of new social problems. Social divisions deepened and crime rates rose. It was also during this period that social concerns started to arise in relation to science, technology and the environment.

During the 1980s there were cuts in public spending. This was also a period of industrial unrest with a number of strikes by trade unions. There were also a range of social and political disturbances throughout this period, and an increasing threat from the IRA. Throughout the 1990s and into the early twenty-first century, we have seen a number of developments in the area of information and communication technologies, alongside an increase in the number of government run databases.

There are a number of implications for the study of CCTV that arise from a focus on the social history of Britain since the end of the Second World War. We can already see from this history that social relations changed substantially during this period. Social inequalities increased at the same time as crime rates beginning to rise. Social disorder also became a prevalent feature of modern society. These changes occurred alongside an increasing political belief in science and technology as representing progress. As I will show later in the book, CCTV is a divisive technology – defined as a technology to protect the law-abiding in society, and to target criminals/'the other'. In the context of the history of Britain during this time, CCTV became a political solution for social problems. The next two chapters will explore the history of CCTV in the context of criminal justice, policing and politics.

Chapter 3
Criminal Justice, Law and Order, and Policing in Post-War Britain

Introduction

> [The police] are not … a straightforward reflex of a burgeoning division of labour … [they may also] develop hand in hand with the development of social inequality and hierarchy. (Reiner 2000, p.5)

In 1829, a 'new police' was formed. The development of a campaign to establish a professional police force began during the late eighteenth century, as various governments came to the realisation that the established systems of crime control and law and order were not enough to combat the problems arising in rapidly growing towns and cities. There was also considerable resistance to the formation of a new police during this time, however this was eventually over-ruled by pressure from the 'urban middle-class propertied interests – seeking a police force to protect their property and persons against what seemed to be an inexorably growing tide of urban crime, and against the threat of revolution by the growing urban masses' (Philips 1985, p.63). The primary function of the new police was to 'prevent crime' (Peel 1829).

There have been a series of major changes in criminal justice, and policing since the end of the Second World War. Crime prevention has been replaced with pre-emptive policing, and criminal justice moved from a focus on rehabilitation to more draconian measures emphasising prison and punitive punishment. Alongside the social and political changes covered in the previous chapter, these developments in policing and criminal justice have focused the attention on protecting the victim, and punishing the offender, often in situations of a divided society in terms of socio-economic status. As the quote at the beginning of this chapter illustrates, it is not just that changes in policing reflect the society around them, but that changes in law enforcement and approaches to tackling crime can also themselves bring about new social divisions and changes in the social hierarchy.

CCTV in Britain did not develop in a vacuum. It is intrinsically tied together with changes in policing and crime control, criminal justice, and political, social and economic factors. In this chapter, I provide a precursor to the next chapter (which concentrates more specifically on the history of CCTV generally), with a history of criminal justice, law and order, and policing in post WWII Britain. In the previous chapter we have seen important shifts in society, government and governance in Britain during the post-war and early twentieth century period, a time of rapid expansion of surveillance and communication technologies. In this chapter, I want to focus on changes in political approaches to law and order, and changes in methods of policing to provide a context for my later analysis of the history of CCTV.

Crime and Criminal Justice

CCTV is often situated in the realm of crime control or prevention, defined as: 'the total of all policies, measures and techniques, outside the boundaries of the criminal justice system, aiming at the reduction of the various kinds of damage caused by acts defined as criminal by the state' (van Dijk 1990, p.205). However, although presented as a tool for crime prevention in political discourse, it is also used as a tool for prosecution and can therefore be said to also be a tool for criminal justice.[1]

After the end of the Second World War there was general agreement, politically, concerning the need to rebuild the economy alongside the creation of a welfare state. This agreement centred on shared goals of increasing employment, a mixed Keynesian economy and a strengthening of social security and an improvement in education and health services. During this time crime and criminal justice matters took a back seat (Downes and Morgan 2007, p.203). During the 1960s these matters began to feature in political party manifestos, due to rising crime rates, although at this time the rise in crime was still not seen as attributable to any political party. It was not until the 1970s that criminal justice policy featured as a major issue in general elections (although it had featured previously but in very minor ways) (Pierre 2006, p.374). Nonetheless, criminal justice policy did exist previously and needs to be looked at in order to gain an idea of the fuller picture of developments in Britain. Furthermore, some argue that criminal justice policy has become increasingly politicised; for example Garland states that changes are based less on trends in crime than with the politics and culture surrounding it (Garland 1993, p.20). He argues that: 'It is clear enough that criminal conduct does not determine the kind of penal actions a society adopts. For one thing, it is not "crime" or even criminological knowledge about crime which affects most policy decisions, but rather the ways in which "the crime problem" is officially perceived and the political positions to which these perceptions give rise' (Garland 1993, p.20). This point is particularly interesting in the context of CCTV. It is a technology used to make the public 'feel safer' – a political position based on enhancing communities' feelings of safety, rather than tackling crime, or the causes of crime, itself.

The rather late convergence of politics and criminal justice is interesting. For some this is due to the 'strength of the belief that crime, like the weather, is beyond political influence; and that the operation of the law and criminal justice should be above it' (Downes and Morgan 2007, pp.201–2). Further to this belief that

1 The criminal justice system is made up of the Home Office, the Police, the courts, and the prison and probation services.

politics has no influence on crime or criminal justice, a rehabilitative approach to the treatment of offenders can also be seen from the 1950s through to the 1970s (Blakemore and Griggs 2007, p.63). In 1948 the amended Criminal Justice Act was introduced, which abolished penal servitude (imprisonment with hard labour) and emphasised corrective training, preventative training and introduced detention centres as an alternative to the harsher borstals used previously (Criminal Justice Act 1948). A year after this enactment, the Royal Commission on Capital Punishment (1949–1953) was set up by the then Home Secretary James Chuter Ede to discuss the possible limitation or modification of capital punishment. The Commission concluded that without overwhelming public support in favour of its abolition, the death penalty should not be abolished (Royal Commission on Capital Punishment 1965).

This focus on rehabilitation in the amended 1948 Criminal Justice Act could be seen once again in the introduction of the Homicide Act in 1957, which restricted the use of the death penalty for murder. Three years later, the minimum age of imprisonment was raised from 15 to 17 under the 1961 Criminal Justice Act. Borstal training rather than imprisonment for offenders under the age of 21 was also encouraged. In 1965, the death penalty was abolished. The Criminal Justice Act was once again amended in 1967, abolishing preventive detention and corporal punishment (Davies et al. 2005, p.xxviii).

During the 1970s, the Conservatives began to make 'law and order' matters more of a political issue, implying that the Labour party was responsible for the rise in crime and arguing in their Conservative Party Manifesto in 1970 that 'the Labour government cannot entirely shrug off responsibility for the current situation' (Downes and Morgan 2007, p.203). The Conservatives began arguing that Labour needed to do more to tackle rising crime rates, placing themselves in the role of what Dunbar and Langdon describe as the 'traditional upholder of equality under the law, in contrast to the Labour party, which was implied to be selective in its view of the law' for the remainder of the decade (Dunbar and Langdon 1998, p.99). The 1979 general election saw the first major post-war difference in Conservative and Labour manifestos on crime issues, as Labour focused on 'dealing with crime through its social and economic programme for tackling inequality, poverty and deprivation' and the Conservatives 'raised the profile of law and order to rank as one of the party's five major tasks' (Dunbar and Langdon 1998, p.100).

At the same time as crime moving onto the political agenda, Garland argues that there was also a break during this time with the idea of 'penal welfarism'. He describes a time in which there was a declining belief in the idea of rehabilitation, a return to using punitive sanctions, and a return of the idea of 'the victim'. He argues that this was surprising as these developments 'involve a sudden and startling reversal of the settled historical pattern … the re-appearance in official policy of punitive sentiments and expressive gestures that appear oddly archaic and downright anti-modern tend to confound that standard social theories of punishment and its historical development' (Garland 2001, p.3). This is the popular interpretation across the literature on criminal justice in Britain. Most theorists agree that policy on crime prevention and criminal justice was severed during the last few decades, with a reversal of what had become the traditional approach to dealing with crime (with restorative and reformative policies taking a central role), to one centred on punitive polices and punishments.[2]

This is certainly the case in criminal justice policy during the 1980s and early 1990s. The 1982 Criminal Justice Act reduced the timeframe of eligibility for parole from 12 to 6 months, as well as replacing borstal training with youth custody (Davies et al. 2005, p.xxviii). However, the 1991 Criminal Justice Act remained consistent with the previous stance on sentencing and punishment, outlining a strategy based on parole, rehabilitation, community sentences and alternatives to incarceration. A couple of years later, Michael Howard (the newly appointed Home Secretary) announced during the 1993 Conservative Party conference that 'prison works'; outlining a new strategy based on rights for victims, loss of rights for criminals, new powers of arrest for the police, and effectively abolishing the right to silence (*The Independent* 19 October 1993). Of the 27 measures outlined by Howard, 18 were included in the amended 1994 Criminal Justice Act (Ferguson 1994).

The criminal justice situation changed fundamentally with the appointment of Michael Howard to position of Home Secretary and the introduction of a far more repressive criminal justice policy (Norris and Armstrong 1998, p.34). At this time there was also a change in the opposition party's stance towards crime and criminal justice. A year before the Conservatives won the 1992 General Election, Tony Blair (as leader of the Labour party and at the 1992 Labour party conference) stated:

> When young men and women seek but do not find any reflection of their hopes in the society around them, when the Tories create a creed of acquisition and place it alongside a culture without opportunity ... [when] people feel they have no chance to improve and nothing to strive for ... [then] in the soil of alienation, crime will take root. (*The Independent* 2 October 1992)

A year later and as Shadow Home Secretary, Blair said:

> There is no excuse for crime. None. (*The Sun* 3 March 1993)

A few years later, during the 1997 general election, the main promise New Labour made to the electorate was one based on crime control (Hoyle and Rose 2001, p.76). Labour had therefore started to take a hard line on crime, marking a real break with its former idea that social factors and underlying causes of crime should

2 For further reading on the development of these methods see Rodman, B. (1968) 'Bentham and the Paradox of Penal Reform' *Journal of the History of Ideas* 29/2 pp.197–210.

be taken into account. The White Paper 'No More Excuses', published in 1997, set out a strategy for the improvement of the youth justice system in combating youth offending. Jack Straw (then Home Secretary) stated in the White Paper that:

> An excuse culture has developed within the youth justice system. It excuses itself for its inefficiency, and too often excuses the young offenders before it, implying that they cannot help their behaviour because of their social circumstances. Rarely are they confronted with their behaviour and helped to take more personal responsibility for their actions. The system allows them to go on wrecking their own lives as well as disrupting their families and communities. This White Paper seeks to draw a line under the past and sets out a new approach to tackling youth crime. It begins the root and branch reform of the youth justice system that the Government promised the public before the Election. It will deliver our Manifesto pledge to halve the time it takes to get persistent young offenders from arrest to sentencing. All those working in the youth justice system must have a principal aim – to prevent offending. (Home Office 1997)

A year later, the 1998 Crime and Disorder Act introduced Anti-Social Behaviour Orders. Furthermore, it provided greater responsibility to local authorities to develop strategies for dealing with crime, in accordance with the police and other local criminal justice authorities. From the early 1990s, there has been an emphasis on crime prevention networks involving state and non-state actors. Garland describes this as an 'enhanced network' of actors, extending the 'formal controls of the criminal justice state' into partnerships, which include more informal methods of crime control. He states that the key phrases used in terms of this strategy included the 'multi-agency approach', 'activating communities', and 'creating active citizens' (Garland 2001, p.125). Whether this happened or happens in practice is another issue though. As an example, more recent policy on CCTV has emphasised a partnership approach to tackling crime. The 1996 White Paper 'Protecting the Public' emphasises a 'co-ordinated approach' to crime control. However, despite emphasising this in theory, the public is often portrayed as a very passive element. I will return to this point later in the book.

Looked at in a broader context, this partnership approach seems to be a general trend in relation to crime prevention in recent years, including the involvement of various partners who are not situated in the traditional realm of criminal justice (Crawford 1998, p.11). Although historically, a range of people held the responsibility for crime prevention, more recently it became seen to be the main function of the police.[3] However, crime prevention is not just carried out by the police, but involves communities, local authorities, technologies, architecture, urban planning and so on.

The 2001 White Paper, 'Policing a New Century', emphasised, amongst other things: 'Better partnership working' (Home Office 2001). A year later, the 2002

3 This point is discussed in greater detail in the next section on policing.

Criminal Justice White Paper, 'Justice for All', once again placed greater emphasis on the rights of the victim, as well as pushing for increased conviction rates, stating: 'this White Paper aims to rebalance the system in favour of victims, witnesses and communities' (Home Office 2002). The previous year, the White Paper, 'Criminal Justice: The Way Ahead', had already proposed: 'putting the needs of victims more at the centre of the CJS [Criminal Justice System]' (Home Office 2001). In the same year, the Anti-terrorism, Crime and Security Act was formally introduced, two months after the terrorist attacks in New York.[4] The Act allowed government departments to collect and share information on terrorist activities, and enhanced police powers to deal with those in custody unwilling to share their identity (Davies 2005, p.xxviii). The 2001 Act was replaced by the 2005 Prevention of Terrorism Act, which allows the Home Secretary the right to apply a control order on any individual suspected to be involved in terrorism.

In this section I have provided an overview of criminal justice in the UK since the end of the Second World War. From the end of the Second World War until the 1970s, rehabilitation and treatment were high on the agenda. At this time, responsibility for crime was moved onto the political agenda by a Conservative Party keen to place the blame for rising crime rates on the Labour Government. From a period of political consensus regarding crime and criminal justice, sprang a divergence of policies and promises, with the Conservatives promising 'law and order' and Labour intent on tackling the social causes of crime. This departure from consensus remained during the 1980s, although rehabilitation, in terms of penal punishment, remained the order of the day. The start of the early 1990s witnessed change with both parties taking a hard line view and stance on crime and criminal justice matters. From this time, a co-ordinated network approach to crime prevention began, with the responsibility for crime prevention falling under the remit of a number of organisations and institutions. Next I turn to the history of policing in Britain, before moving on to a more concentrated history of CCTV (and its use by the police, as well as a more general social history).

A Short History of the Police from the 1950s

In this section I introduce a concise history of the police in Britain, from the 1950s onwards. I am not aiming to cover all bases here but rather to give a general idea of some of the principal shifts in policing that occurred during this time. This is important for situating the later analysis of the use of CCTV by the police during this time, in the wider context of changes in policing (and furthermore, wider changes in society and politics during this time).

---

4 The terrorist attacks in New York took place on 11 September 2001. They were a series of co-ordinated suicide attacks by al-Qaeda (a Sunni Islamist terrorist group) on a number of buildings in the United States.

During the 1950s, policing remained largely organised around a beat system, with the police in rural areas for patrolling entire areas, and those in urban areas divided into shifts (Newburn 2005, p.85). A move towards motor vehicles for the police was strongly rejected at this time. The Royal Commission cited a variety of disadvantages of moving from a beat to motorised system of patrol, including the argument that it would 'diminish that contact with the public which is so useful to the police and to the public itself'. Affection towards beat policing, both within and outside the police, remained strong well into the 1960s (Newburn 2005, p.85).

Over 70,000 police officers were employed in Britain by 1959. The 1960 Commission, set up to review police pay and administrative issues, recommended appointing a Chief Inspector of Constabulary responsible for scientific research, development and planning. For some, this development 'paved the way for ... more technopolicing' (Manwaring-White 1983, pp.20–22).[5] Radio had already been used since 1951 to increase efficiency, and criminal record offices were being set up to allow a greater flow of information between forces.[6] The telephone was also being used to a greater degree. Manwaring-White argues that in terms of these technological developments:

> No checks or balances however were advocated by which the spread of this new scientific approach to policing could be evaluated. The dangers to the fragile freedoms of civil liberties, posed by sophisticated eavesdropping, electronic surveillance and computer information gathering were not considered. Technopolicing was embraced wholeheartedly as the new scientific aid to better policing. (1983, p.22)

The 1960 Commission recognised the importance of the British public trusting and having confidence in the police, referring to this as the 'British police advantage', heralding the relationship as a unique one. However, this relationship was to change during the 1960s. As I mentioned in the previous chapter, society started to develop in an insular way, with the growth of high-rise buildings and major changes in community interactions and relationships. This was coupled with a new system of reducing the number of police officers patrolling areas on foot, and instead putting them into cars, which became known as 'unit beat policing'. A Home Office circular released in 1967 encouraged this system, stating that its main advantage was coverage of wider geographical areas over a 24-hour basis (personal radios were also being provided to police officers at this time). A system designed to improve the amount of areas covered turned out to have a detrimental effect on police-community relations, due to a concentration by police officers on the 'action' element of their role, rather than the 'service' element (Newburn 2005, p.85). Chibnall (1977, p.71) argues that:

5 Manwaring-White defines technopolicing as the 'technology of policing' and the 'science of policing and its tools'.

6 I am referring to telephony here. Telegraphy had been used since about 1934.

> The dominant image of the honest, brave, dependable, (but plodding) 'British Bobby' was recast as the tough, dashing, formidable, (but still brave and honest) 'Crime-Buster'.

Bradley et al. (1986, p.16) reiterate this point and argue that;

> The introduction and spread of the unit beat system in the 1960s is now regarded as at worst an unmitigated disaster, which hastened the growing divorce between the police and the public, at best a desperate stop gap form of fire-brigade policing.[7]

Furthermore, during the 1950s and 1960s, a number of regional criminal record offices were set up with the aim of keeping a daily record of all information relevant to a crime, as well as a manual index of suspects and criminals in the local area. Records were also held on disqualified drivers and those with drug convictions, as well as a collation of information of criminal fraternity in each area, beginning what is termed by Manwaring-White as 'pre-emptive policing'; acquiring information on a person before they had committed a crime. The advent of computerisation made this record keeping far less time consuming, however as Manwaring-White (1983, p.54) argues little thought was given to the privacy impact of this new method of data collection or to the issue of restrictions in terms of access to the information held.

Rawlings (2002, p.201) argues that at this time, the 'general climate of fear created by the Cold War' together with the start of mass protests against nuclear weapons and other protests occurring during the 1950s, 'made the police and the government nervous about dissent'. This, coupled with an 'increasingly critical outlook by a more widely and highly educated public', saw the relationship between the police and the public suffer during this time (ibid. p.201). However, the Royal Commission did not acknowledge that there were any problems between the police and the public (ibid. p.203).

The early 1960s saw a record number of criminal offences being recorded. In response to these rising crime rates, the National Crime Prevention Centre was established in 1963. The centre functioned as a national police training unit, set up as a result of a realisation on the part of the Home Office that the police required 'active cooperation' from the public if crime was to be prevented (Gilling 1993, p.233). This in turn meant that the Home Office required the 'cooperation of the police to get the message across to the public' (ibid.). Following this, the Cornish Committee on the Prevention and Detection of Crime was set up, with the purpose of making recommendations to the Home Office o the subject of crime prevention and detection (Crawford 1996, p.26). Although crime prevention had been thought of as a key function of the new police since the beginning, what this actually

7 Fire-brigade policing was the term given to the police responding to crime, rather than seeking to prevent it.

meant in practice had been unclear (Newburn 2005, p.87). Following the 1965 report from the Committee on the Prevention and Detection of Crime, a number of specialist crime departments were set up. Until the early 1980s, responsibility for crime prevention remained under the control of these specialist departments, treated as 'a peripheral specialism of low status and interest when placed alongside crime-fighting' (ibid.). However, the continuing rise in crime during the 1980s meant that 'one of the key messages emanating from the police was they could not be expected to carry responsibility for the prevention of crime unaided' (ibid.).

The 1964 Police Act saw a shift in power away from local to central control, with the Home Office gaining influence in terms of determining police policy and administration. The Act also merged smaller police forces together with larger ones (Rawlings 2002, pp.204–5). In 1969 the Serious Crime Squad was established on a permanent basis. Policing at this time was therefore becoming more centralised, and controlled to a greater extent by the Home Office.

The 1970s saw the advent of the use of information technology by the police. In 1972, the first computerised command and control system (computerised system for despatch of resources) was put into place by the Birmingham police force (State Research Pamphlet no.2 1981, p.4). A more advanced command and control system was implemented in Glasgow during 1973, and by the end of the year the Home Office had issued a memorandum of guidance to all chief constables advocating the potential benefits of the system to all police forces (ibid.). Subsequently, in 1974, the Police National Computer (PNC) was created, consisting of a database for stolen vehicles. Since its inception, more applications and databases have been added.[8] By 1977, forces in the West Midlands, Strathclyde, Staffordshire and Suffolk had set up command and control schemes, followed by a number of other areas. In 1979, the Metropolitan and City forces jointly ordered the largest computerised command and control system in Britain (ibid.). John Alderson, Chief Constable of Devon and Cornwall at the time, stated:

> The modern generation of police officers are beginning to see themselves as mobile responders to incidents. The car, radio and the computer dominate the police scene. The era of preventive policing is phasing out in favour of responsive or reactive police. The main casualty of this change is police/public understanding, confidence, and in some cases, trust. Police more often meet people in conflict or stress situations and the opportunities to strengthen police/public understanding in a day-to-day fashion diminish. The technological cop has arrived. (Alderson 1978)

8 By 1983, the PNC held the details of over 30 million vehicles in the Vehicle Owners File, fingerprints were held on the Fingerprints File, and there was a broadcast system, which could be used by a police force to send messages to any other force or group of forces. Currently, the Police National Computer holds details of people, crime, vehicles and property; accessible to the police and other criminal justice agencies. It is managed by the National Policing Improvement Agency.

Emsley (1995) describes the rise in crime during the 1970s and 1980s as prompting major criticism of this type of reactive or fire-brigade policing. However, he argues that at the time the shift from on the beat policing to one based on police in motorised vehicles was greeted with enthusiasm by the media and politicians. However, some others did not greet it with the same enthusiasm. For example, Statewatch (1981, pp.3–4) argued:

> It is a system of policing which, because it places efficiency at the forefront, not only leads the police in conflict, but necessarily negates 'community/preventive' policing in any meaningful way. It also relies on an ideology which designates part of inner cities as 'high crime areas' – those working class areas of high social deprivation and often large black communities like Brixton, Hackney and Lewisham in London, Huyton in Liverpool, and Lozells in Birmingham – where policing is not a question of protecting the community but of keeping it under control.

After the Conservative government came to power in 1979, there were increased resources provided to the police in order to bring down crime rates. However, by the end of the first term rising crime rates and falling detection rates showed the strategy of increasing resources in order to decrease crime had not had the desired outcome (Rawlings 2002, p.213). The end of the 1970s and throughout the 1980s saw a rise in public disorder, coupled with a 'creeping crisis of confidence in the police', which resulted in a 'succession of competing agendas for reform' to emerge (the Conservatives wanted to give greater powers to the police whilst Labour pushed for greater accountability) (Reiner 2005, p.689).

The Scarman Report was published in 1981, in the aftermath of the Brixton riots. Some have suggested that the cause of the disturbances in the 1970s and 1980s were a direct result of public alienation from the police (see for example, Ackroyd 1993, p.5). Others have argued that the riots were a protest against society, by those who felt socio-economically deprived. The Scarman report suggests that either explanation is an over-simplification of a highly complex situation. The conclusions of the report stated that there was:

> i) the need for a concerted, better co-ordinated attack on the problems of the inner city; ii) recognition of and action to meet the special problems and needs of the ethnic minorities, based on an acceptance of them as full and equal members of a culturally diverse society; iii) the need to involve not just black people, but all the community, both nationally and locally, in a better directed response to these problems. It is essential that people are encouraged to secure a stake in, feel a pride in and have a sense of responsibility for their own area; iv) the role of the police as essential participants in any effective response by the community (1982, p.14).

The Scarman report led to a number of reforms in police management techniques; however the eventual outcome was not a positive one. The late 1980s saw 'all-time record crime increases, renewed public disorder, spectacular scandals involving miscarriages of justice and plummeting public confidence in the police' (Reiner 2005, p.690). Ackroyd argues that the underlying cause of the lack of public confidence in the police was the take up of new technology, its impact on the organisation, and a managerial failure on the part of the police to deal with these developments (Ackroyd 1993).

The 1980s also saw the rise of community policing and inter- or multi-agency cooperation, as well as a change in the role of the police, such as the involvement of the Metropolitan Police in the Brixton Riots of 1981 (Newburn 2005, p.87). In 1983, Sir Kenneth Leslie Newman (appointed as Commissioner of the Metropolitan Police in 1982) developed a new statement of the Principles of Policing, redirecting a number of the primary objectives of policing developed in 1829. Newman's strategy was based on the idea that the public must be involved in order to combat crime successfully. He extended the idea of policing by consent to one involving a 'notional social contract' between the public and the police; which was to be realised through greater involvement of the public in consultative committees and crime prevention panels (Brake and Hale 1992, p.75).

The 1980s also witnessed the birth of Neighbourhood Watch schemes. Emsley (1995, p.141) describes these schemes as 'link[ing] back to traditional, informal modes of community behaviour', which 'encourage[s] watchfulness and suspicion'. He goes on to suggest that the schemes can be seen as a failure of preventive policing and argues 'by their formal creation, the police are implicitly admitting that, after years of insisting they were the experts in crime prevention and detection, they cannot solve the problem of crime alone'.

The Police and Criminal Evidence Act was set up in 1984; a legislative framework governing the powers and duties of the police, and providing codes of practice for the exercise of those powers. The aim of the Police and Criminal Evidence Act was stated as being to balance the powers of the police to bring offenders to justice with the rights of the public (Zander 1990, p.vii). In 1987, when Sir Peter Imbert took over as Commissioner, he continued to emphasise the need for police-public cooperation and consultation.

In 1994, the objectives of the Metropolitan Police were designed for the first time by government. The 1994 Police and Magistrates' Court Act legislated that the Home Secretary was allowed to issue Codes of Practice, which 'regulate the manner in which police authority pursues policy objectives' (Barnett 2011, p.717). During the same year, the Home Office and Association of Chief Police Officers issued the National Strategy for Police Information Systems, essentially an information technology strategy (Hoey 1998, p.69). The 1996 Police Act extended the powers of the Home Secretary, allowing them to 'make regulations for the government, administration and conditions of service of police forces' (Barnett 2011, p.718).

As mentioned in chapter 1, anti-terrorism has become a major part of the police role during the early twenty-first century. The extended powers of the police to detain passengers at airports and other transport hubs for up to seven hours without the need for any prior 'reasonable suspicion' that someone is involved with terrorism has provoked criticism from human rights groups, such as Liberty. A recent response from StopWatch (a coalition of academics, civil liberties campaigners, legal experts and campaigners) to an HMIC report on Stop and Search, cited a general increase in Stop and Search powers as 'causing damage between police and communities across the UK' and that 'the lack of consent and respect felt by young people during those encounters undermines faith in policing, undermines policing by consent and even contributes to the development of public disorder' (StopWatch 9 July 2013).

In this section, I have described the changing nature of the role of the police in the latter half of the twentieth century. I have shown that crime prevention has taken a back seat at times to other responsibilities, such as responsive or reactive policing. I have also described the introduction of new technologies and societal changes occurring in the 1960s; a withdrawal of society and decrease in on the beat policing. The detrimental effect on police-community relations and an increased focused on 'crime reaction' rather than prevention, coupled with this decrease in on the beat policing is one possibility for the use of CCTV to fill in the gap in police on the street, which will be looked at in the next section. During the 1980s, the idea that the police could not tackle crime alone meant that there was another possible place for CCTV as a tool for crime prevention, which again will be explored in the next section. In light of further changes in information technology, technological advances, and a push for the police to 'fight crime', alongside a lack of confidence on the part of the public and rising crime, we will next focus on where CCTV could potentially fit in as a technological solution to social problems. In the next section, I look more specifically at the use of CCTV by the police from the 1950s.

Police Use of CCTV up to 1984

This section draws on empirical evidence gathered from the Association of Chief Police Officers (ACPO) archive, held at the Open University (which includes letters, articles and details of exhibitions), as well as media articles and academic sources, in order to piece together the history of the use of CCTV by the police from the 1950s to 1984. I concentrate most of the detail of this section on the empirical evidence from the 1950s and 1960s, as this is the most under-researched area of police history with regard to video surveillance.

In 1959, two newspaper articles (*The Times*, 30 June 1959; and *The Times*, 16 November 1959) report the use of CCTV for traffic control and monitoring purposes. This use of CCTV for road traffic control is described as occurring for the first time, anywhere in the world, in Durham in 1959. However, archival research, using the ACPO files from the 1950s reveal that experiments with cameras for traffic management were conducted already being conducted a few years earlier. An article from the *Police Review* included in the ACPO files detail experiments being conducted with CCTV in August 1956, in Durham (ACPO *Police Review*, 11 August 1956):

> Durham experiments with Television – Experiments in the use of television are to be made in Durham next week. Cameras attached above two bridges will be linked to a 5-in screen in a shelter from which a Police Officer directs traffic. The picture on the screen will show the Policeman how much traffic is waiting to enter the narrow streets leading to the market place. It is believed to be the first occasion on which television has been used for traffic control in this country.

At the beginning of May 1957, the Lancashire Constabulary gave a demonstration of the uses of CCTV for traffic control and management, by installing cameras onto four main roads and allowing participants to view the images taken. This demonstration was described as the largest of its kind and was in partnership with a number of radio and television manufacturers (ACPO *Police Review*, 7 May 1957):

> Television Experiment – A demonstration of the way in which television can be used to help in the control of traffic was given at Lancaster during the Easter week-end. It is believed to be the biggest demonstration of its kind ever staged in Europe, and it was arranged by the Lancashire Constabulary, in co-operation with a leading firm of radio and television manufacturers … The Durham Constabulary and the Hamburg Police have already shown what television can do to help the Police to regulate the flow of traffic at an intersection where the immediate approach roads are hidden from the pointsmen on duty. The Lancaster experiment was on a much more ambitious scale. The area included four main roads … These roads were all observed by closed-circuit television cameras.

The ACPO files also detail the Minister of Transport approving a permanent installation of CCTV cameras in Durham Market Square; also stating that he had agreed to contribute towards the cost (ACPO *Police Review*, 6 July 1957).

A letter from a manufacturer to ACPO in late 1958 described the potential uses of CCTV as being for: 'road safety, teaching crime prevention to the public, various training uses within police departments, and recording and combating dangerous driving' (Rank Precision Industries letter, 12 December 1958). The response from the Secretary of ACPO, N.W. Goodchild, stated that: 'I am particularly interested in film making for instructional and propaganda purposes under the headings you mentioned' (Response from Goodchild, 22 December 1958). A letter in the archives dated July 1959 from Burgot Rentals Ltd. to all Chief Constables detailed two meetings to be held at Aldermanbury House for crime prevention officers (on 4 and 5 June, 2–6pm) to 'show crime prevention officers a collection of exhibits and working models that are available for use as crime prevention exhibitions' (Correspondence with Burgot Rentals Ltd. 4 May 1959). In July 1959, a letter from Goodchild to all Chief Constables, described plans for a conference exhibiting a range of equipment under exhibit categories. This collection of exhibits included radio, radar, telephones, emergency communications equipment, and television, collected from a variety of police forces (Letter from Goodchild, 30 July 1959) For devices used in the detection of crime, Sheffield police force offered a 'burgot cabinet camera'. For traffic control purposes, the Metropolitan Police offered 'raised control posts' over the Chiswick flyover (ACPO PEE schedule of exhibits in the letter from Goodchild, 30 July 1959). From the point of view of some police forces, there was not a great deal of enthusiasm for CCTV during this time, despite a 'hard sell' from industry, including one letter which described recent improvements in CCTV technology (this appears to be a circular letter sent to every Chief Constable):

> Dear Sirs,
>
> Closed Circuit Television
>
> Even with the present fast expanding interest in EMI Closed Circuit Television equipment, it is still not often realised how recent advances in Technical Performance have improved sensitivity, resolution and reliability, and that an installation consisting of a Camera, Control Unit, Camera Tube, Len and Cable can cost less than £550. There may well be in your own particular sphere problems which can be solved or improvements made by the use of such equipment in Training Methods, Traffic Control or Security Work, etc. We would therefore like to offer, at a date and time convenient to yourselves, a visit from one of our Engineers to discuss ways in which Television could be of assistance. In the meantime, we trust the enclosed details will be of interest and look forward to the opportunity of assisting you further. (Letter from DG Ashton Davies, Sales Manager, Instrument Division, EMI Electronics Ltd., 7 October 1959 – emphasis in original)

A handwritten note on the back of this letter, signed by the 'Inspector, Traffic Department' stated: 'Sir, This is just another circular to bring to notice the latest developments in television which might be of use to us. I feel a visit and demonstration is unnecessary and suggest we file' (Handwritten note, Inspector Nichols, 8 October 1959). With this covering letter, EMI included a glossy brochure, stamped with Hertfordshire Constabulary, 8 October 1959. Inside they described CCTV as:

> One of the most versatile aids to management. It can help increase efficiency and profit in many ways. Here are just a few:

> Finger-tip control – several operations, at widely separated locations, can be co-ordinated by remote inspection. Hazardous operation – processes too dangerous to be inspected on the spot can be seen quite clearly in perfect safety. Training – large audiences of students can be given a close-up view of experts in action. Sales promotion – demonstrations can be projected throughout a store or on to an exhibition stand: sales conferences can be addressed by directors from distant offices. Data transmission – blue-prints and other visual data can be verified from a distance: signatures can be verified. Surveillance – traffic can be controlled: fire and theft can be detected from key positions in factories and warehouse. [sic] Microscopy – microscopic phenomena can be viewed simultaneously by large audiences. Ultra-violet and infra-red – even that which is invisible to the naked eye can often be seen on television.
>
> Closed Circuit television is already useful in hundreds of different ways. There may well be yet another application, to solve your particular problem.

For industry at this time, there were a variety of applications for CCTV (only one of which is surveillance-related, and even then only for traffic control, or safety purposes, rather than surveillance of public spaces and so on).

In 1960, at the ACPO No.6 District Conference, under a set of minutes entitled 'Closed Circuit Television use by the police' the outcome of an experiment with CCTV was described as follows:

> Mr. [surname anonymised] referred to a visit which he had made to a recent exhibition of television equipment. There, he had been impressed with the use which could be made of closed circuit television. It appeared, he said, that it could be used for the transmission of fingerprints and photographs from what he had been able to ascertain the running cost might be cheaper than Talex [sic]. When at the exhibition he had been given to understand that the representatives from the Metropolitan Police and City of London Police had visited there and examined this equipment … Mr [surname anonymised] said that closed circuit television had been used during the course of a recent royal visit. A camera had been put at certain places on the route where, it was thought, difficulties might arise in controlling traffic in crowds. Following this experiment, it was considered that there had been no special advantages gained. Here Mr. [surname anonymised] said that he was particularly interested in the transmission of photographs and such material. He asked if such a transmission could be copied. The Chairman said that the use of closed circuit television had been considered by the Wireless Sub-Committee, the main difficulty was the cost … He understood that closed circuit television had proved its usefulness in large offices and banks, but in no way could it be regarded as an eventual substitute for the Telex system.

It is likely that the 'royal visit' referred to was a state visit by the Thai royal family to London, during which EMI lent the Metropolitan Police two CCTV cameras to put up in Trafalgar Square (Williams 2003, p.10). Once again, during this time, CCTV was portrayed as a solution by industry but when it was actually used there was a fairly unenthusiastic response from the police. However, the cameras were used again later in the year to monitor 'the usual Guy Fawkes rabble' in Trafalgar Square on 5 November (ibid. p.13). However, Williams described the trial use of these cameras as a failure (due to rain and inadequate lighting) and cites a policeman at the time as saying: 'in dealing with ceremonials, potentially disorderly meetings and the like I know of no substitute for the experienced police eye on the spot and the ability to sense trouble in the air'. (ibid. p.13)

During the 1960s, the police installed CCTV cameras at a variety of locations for traffic control purposes. One report from the Chief Constable of Kent detailed the installation of cameras in the Dartford Tunnel in 1964. This installation consisted of a number of cameras (the report doesn't detail the exact number but it seems to be at least 6 cameras) and two monitors, along the roof of the tunnel and at both entrances. The report stated that the installation was 'provided under the direction of the Ministry of Transport' and that 'it was originally envisaged that it would only be subjected to casual use'. However, the report goes on to say that:

> 'Operationally it has been found necessary to have the installation in use almost continuously'. The reason for this continuous use is a technical issue: it takes 23 seconds to switch from one camera to being able to view a picture from another (described as a 'cold camera'), so all cameras remain switched on all the time to avoid this problem. There is an additional problem in terms of picture quality, and although there is a description of being able to follow a vehicle through the tunnel, the report also states that 'it is not possible to read the registration number of a vehicle going through and it is often not possible to identify the make of a vehicle'. (Chief Constable of Kent 1964)

Although there are various references to CCTV being used for the first time in November 1978 (according to newspaper sources that various theorists use) to catch criminals (see for example, McGrath 2004; Moran in Norris et al. 1998), I have found examples that are much earlier than this. For example, in November 1964, *The Times* reported the installation of CCTV systems throughout Liverpool city centre 'to watch certain spots where crime is likely'. The four cameras, lent to the police by electronics companies, were placed in a number of locations in the central business district (Williams 2003, p.6). Although the initiative was described as an 'experiment in the fight against crime' (*The Times* 26 November 1964) the practical application of the cameras did not seem to add anything different to simply having a policeman with a radio looking out of the window. The images were being transmitted to monitors in the same buildings as the cameras were mounted. For Williams, this was undoubtedly a PR exercise, and I would agree. Publicity at the time focused on the potential for CCTV to detect and fight crime, and a positive reaction from the public was reported (Williams 2003, p.6). Liverpool police depicted the experiment as a success. However, the figures reported were based on a three-week period, at the time of media reportage of the experiment (*The Times* 26 November 1964). Publicity surrounding the experiment may also have produced a deterrent effect, which was noted by Home Office statisticians when assessing the initiative. The Home Office at this time were far more sceptical about the results produced. Eventually, even the Liverpool police admitted that the effect was preventative but most likely due to a psychological effect (deterrent effect) (Williams 2003, p.6). Three months after the initial experiment, an article described the potential for these CCTV systems to be coupled with infra-red technology in order for them to be able to be used in the dark (*The Times* 17 February 1965).

In 1968, *The Times* once again reported CCTV being used for crime prevention and detection purposes, with:

> The first experiment in London in the use of closed circuit television to detect and prevent crime has been started in Croydon by the Metropolitan Police … Unknown to the local public, the Croydon experiment has been going on since December 18 … The cameras are controlled from a small room in Croydon police station by two operators who are also in radio contact with police patrolling the area … The police believe the main use of the cameras will be in detecting car thieves and pickpockets … Asked about possible criticism that the scheme had been started without the knowledge of the public, he [Chief-superintendent J. Hall] said: 'What is the difference between being looked at by a man in plain clothes whom the public do not recognize as a policeman, and a camera they do not see?' (*The Times* 4 January 1968)

This quote is particularly interesting as it shows ownership of the technology as seen by the police. The public were not involved in decisions over CCTV, and the cameras are described as having been installed for crime prevention purposes, utilised, installed, controlled, and owned, by the police. Perhaps even more importantly, CCTV is constructed in this instance as a simple replacement for a police officer. By asking what the difference is between a police officer and a camera (and answering that there is none) there was an attempt to justify the use of covert surveillance. This line of questioning essentially implies that there are no other issues to be resolved in the use of a technological system (rather than using human intervention). So, for example, there seems to be no difference in perceptions of accountability.

In 1968, a temporary camera was installed in Grosvenor Square for the purpose of monitoring any-Vietnam protestors (a previous protest had resulted in a high level of public disorder). Following this, in 1969, a number of permanent cameras were installed in Grosvenor Square, Whitehall and Parliament Square. This was the first permanent installation of a CCTV system; installed to 'protect the heart of the state' (Williams 2003, p.16).

Although the technology was now more readily available to the mass market, Norris and Armstrong (1999, p.35) argue that:

> The political climate retarded its introduction. The early 1970s and early 1980s saw a fierce political confrontation between elements of the local state, particularly Labour controlled local authorities, and the police regarding accountability.

Further to this, Williams suggests that there were also other reasons that the uptake of CCTV was slow for the police; for example, the cost of the actual cabling needed to transmit images was too high (2003, p.15).

CCTV was installed during the mid-1970s at Clapham Common, Stockwell, Clapham North and Brixton underground stations (Mayhew 1979, pp.21–3). However, police use of the technology remained for the purpose of traffic control in the majority. I argue that this was the case until the 1984 Miners' Strike with the introduction of CCTV cameras in residential areas, to monitor the movement of miners and residents away from the pit sites. To my knowledge this is the first use of CCTV to surveille a residential area. Although it has been suggested that the use of CCTV during the 1970s and early 1980s 'remained limited and focussed on marginal groups such as football hooligans and political demonstrators' (McCahill and Norris 2002), the extracts from diaries of miners show that CCTV was also installed throughout residential areas, and although not installed as a permanent installation for surveillance of public areas, the strike lasted a year, with surveillance of both miners and other residents. It was also during this time, I argue that CCTV was defined as a political tool and one of surveillance.

Surveillance during the Miners' Strike

During the miner's strike the police and intelligence agencies employed numerous methods of surveillance. The Home Office had assured the Association of Chief Police Officers that resources for policing the strike would not be a problem (Marsh et al. 2004, p.145). Furthermore, the Home Office announced that the police had the right to photograph miners in custody without their consent, providing they did not employ force (Armstrong and Giulianotti 1998, p.122). However, according to reports from miners, photographs were taken without consent at roadblocks, at which time vehicle registration details were also taken (WCCPL 1985, p.59). These vehicle registration details were then entered into the Police National Computer at Hendon, in order to keep records of cars of interest to the police (East and Thomas 1985, p.140). Miners recalled the experience:

> My car has been checked when we were picketing in Leicestershire … we were outside Bagworth colliery … we were outside picketing … nothing was going on. I was standing there and one policeman came walking up along the side, looking at my car, my tyres, my tax and looking at my number plate, and he radioed it through on his radio. You could see him looking at my number plate

> and speaking straight through on his radio ... He was telling them the number of my car. Other pickets had their cars checked and all. All the pickets' cars were parked in one line and this policeman went up past them and we tackled him about it. He said 'I'm entitled to do it. I'm checking the cars'. And he checked the number and radioed them through on his radio.
>
> On the way to Leicester our car was stopped by the police and we were threatened with arrest. They took the number of our car. (ibid. p.60)

Reports of photographs at picket lines and video surveillance from miners are also numerous. The official position from David East, then Chief Constable of Wales, was as follows:

> It is not the usual practice of this Constabulary to take photographs at demonstrations. However, where there is good reason to anticipate a breach of the peace or the commission of other offences, the police are justified in obtaining photographic evidence. Film is sometimes taken for the training of officers in crowd control techniques. Of course, many people, including the media, take photographs on such occasions. (ibid. p.139)

The miners involved describe the situation as follows:

> At Margam there was a photographer wandering around taking our pictures. Later we saw him at the police canteen eating sandwiches and drinking tea. He was laughing and joking with an inspector as if they had known each other for years.
>
> At Margam video cameras were outside the gates. One was put in the bedroom of a private house, another in a Range Rover and another was seen on the hard shoulder of the M4.
>
> At Port Talbot they had a video on us. At Bagworth Colliery in Leicester on top of the canteen they had a video going around all the time, watching you all the time.
>
> In London, on the lobby of Parliament, we saw a police helicopter. We also saw police on office blocks with cameras with telephoto lenses. (ibid. p.139)

The use of video surveillance is also mentioned in a diary extract (from Iris Preston, a member of one of the mining communities) detailing the records of a 'discussion in the community' meeting on the 2 March 1985:

> The camera. This is now being used at nearly every pit and is known as 'big brother'. The miners resent it. Even some scabs – they feel it is another intrusion

> and restriction in their lives. They were not sure whether it recorded continuous movement in the pit area and put it on film to be used against them. The women felt it could take photos of the homes of miners near to the pit. This pit has a public path running through it from one village to another. The camera restricted people from using this path. The old people in the village expressed a little fear of it. Most of the community was disgusted by the use of this camera and felt it was not just there to protect NCB property as the complement of police was heavy and not much damage was done at this pit. (Bloomfield et al. 1986, pp.114–15)

The use of the video camera is mentioned again further in the diary, alongside comments on the general nature of policing during the strike:

> There are also constant police patrols, NCB security patrols. Video cameras fitted with intensifiers for use in the dark have been installed on the winding headgear. Resentment of the police presence on the tip is voiced by the strikers everyday … The brutality of the riot police drafted in from Southern Counties has left deep scars on the minds of the people of this village, and an inheritance of hatred and mistrust of the local police. This hatred will take until long after the strike is over to heal – not months, but generations. (ibid. p.121)

Another diary extract, this time from a pit delegate, details the technical nature and capabilities of the video cameras used:

> The next report involves our meeting with Deeming and his new razor sharp deputy manager … a man who has been sent in 'to deal with us'. This meeting took place in the one time 'Special Branch Surveillance Room' which had just been cleared of its spy cameras and snatch squad devices. Incidentally we accidentally came across this equipment earlier and its power was far greater than we had thought. From the pit its trajectory was 360° in all directions (including up and down) the most chilling thing however was its magnification. From the pit it could focus on every street on Broadway, individual houses, it could pick up clear details in back gardens in Hatfield and of course Stainforth had complete coverage. The Snatch Squad raids on individuals in the pit lane were obviously co-ordinated from here, as were the police attacks up certain streets and the filth laying in waiting on Broadway. (ibid. pp.228–9)

CCTV was therefore used in an intrusive fashion, to surveille residential areas. It was not only picketers that were targeted at the sites, rather the cameras were also able to look into private households in residential areas surrounding the sites.

Surveillance of the strike is interesting as it is an example of (and time of) the convergence of surveillance and criminal justice. Surveillance in this context was used for both exclusionary purposes, due to political reasons, as well as, to my knowledge, for the first time to surveille residential areas (therefore moving towards a mass, rather than targeted surveillance). Video surveillance was used to restrict access to strike areas; as well as for control purposes once the miners were assembled. However, it was not only miners who were affected. There are also reports of the alienation of residents of the mining towns, hindered from going about their everyday lives due to the appearance of surveillance cameras in the towns. Furthermore, other citizens stopped at roadblocks reported a sense of alienation due to police surveillance at access points to mining villages and towns. Although the use of CCTV did not immediately and dramatically increase in terms of numbers (and therefore did remain limited as Norris and Armstrong suggest, above), I believe the use of the technology during the strike to be a major factor in defining and constructing CCTV as a political tool, utilised to attempt to control and marginalise 'problem' groups in society.

Some theorists argue that the police have always been cautious when it comes to new technologies (e.g. Goold 2004), and this certainly seems to have been the case with CCTV. Although the police experimented with the use of CCTV for traffic control purposes as early as the 1950s, they seemed reluctant to utilise CCTV for the wide variety of possible uses proposed by industry during this time. I tend to disagree with Goold's idea of how the police have approached new technologies in general though. As an example, the police advanced the state of the art in radio during the 1920s, 1930s, 1950s and 1960s; they were the first to use a real-time computer; and they adopted the teleprinter in the 1930s (it might be argued too early, rather than too late). So, the rather cautious attitude found in the previous empirical evidence is interesting. It seems to stand in contrast to how the police have approached new technologies more generally.

Conclusion

The archival evidence in this chapter shows a general reluctance or lack of enthusiasm from the police to take up and utilise CCTV when it was first developed. Instead we see a push from industry, citing a variety of benefits of this new technology, one of which was cited to be its capacity for surveillance (although this is not often the main emphasis). The technology at this time was constructed as a multi-purpose system that could be used for training, health and safety purposes and for transmission). Surveillance of the public via CCTV was not an idea implemented immediately by the police. Instead it was a political event (the 1984 Miners' Strike), which prompted the first use of CCTV by the police in a residential area (although as mentioned previously there had been experiments with CCTV in public space prior to this time).

This move towards CCTV being used as a tool for surveillance and politics occurred alongside a variety of changes in policing and in society. As described in the previous chapter, it was not only the police that got into cars, but also the public. Society began to withdraw. This withdrawal of society occurred alongside a rise in relative poverty. Although there was a rise in wages and a fall in the price of consumer goods, the gap between those becoming more affluent and those with little widened and became more pronounced throughout the 1970s (Marwick 1996, pp.224–5). As mentioned in the previous chapter, housing estates, which had been built to tackle housing problems, had become areas for social decline and rising crime.

It was during this period that criminal justice and the problems of rising crime rates became a political issue. During the 1950s and 1960s, criminal justice had focused on corrective training and rehabilitation. There was a genuine shift in approaches to criminal justice during the 1970s. The way in which crime started to become constructed also started to define the solution. The notion of criminality started to become synonymous with 'the other' in society – a group to be targeted and demonised, rather than looking at the social causes for crime. During this time, policing methods also changed. During the 1950s police officers had operated on a beat system, maintaining contact with the public and good relations. The next decade saw a move towards increasing use of technology by the police, as well as a distinct change in their role, towards that of a 'crime buster'. Alongside rising crime rates, there was also the evolvement of 'pre-emptive' policing – collecting information on members of the population prior to any crime being committed. During this time the Home Office also started to become more involved with formulating police practices. Coinciding with a move towards more draconian measures of punishment for those involved in criminal activity, the police became mobile responders to incidents, using an increasing number of technological means in their everyday activities and approaches to tackling crime. The public became more distanced from the police and police-public relations continued to diminish.

The 1960s saw the start of the use of CCTV by the police. During this period, CCTV was experimented with for a variety of purposes – traffic control and to tackle public disorder. However, a number of technical issues, as well as the problem of expensive cabling, meant that it was not used particularly widely. From the ACPO files we can see a general lack of enthusiasm during this time towards the various proposed uses cited by industry for the technology. The police and Home Office also seemed to be at odds as to how to use the technology. During the late 1960s, the discourse from the police surrounding the technology was that it prompted no cause for concern – it was cited as being no different to a policeman on the street. In addition to the differing expectations of CCTV from the Home Office and the police, and the cost of cabling, crime had not yet reached the political agenda during this time.

During the 1970s and 1980s, the issue of rising crime rates became a political issue. CCTV started to be used for the purposes of tackling crime and social disorder in the London Underground, and a number of football grounds (this will also be looked at in more detail in the next chapter). The 1990s saw an even greater shift in criminal justice policy towards the idea of 'the victim' with proclamations from politicians that 'prison works'. This idea of the need to protect the innocent, and incarcerate the criminal, and that this was the solution, presented the perfect opportunity to utilise CCTV as a political tool to be seen to be doing something about crime. Constructed as something to provide safety, it often becomes justified in terms of its role in prosecuting offenders (who are then incarcerated, and according to political discourse 'prison works'). In the next chapter, I continue with a more general history of CCTV.

Chapter 4
A History of CCTV in Britain

Introduction

> The huge rise in surveillance and data collection by the state and other organisations risks undermining the long-standing traditions of privacy and individual freedom which are vital for democracy ... If the public are to trust that information about them is not being improperly used there should be much more openness about what data is collected, by whom and how it is used. (Lord Goodlad 2009)

The above quote is included in the 2009 House of Lords Committee report on the surveillance society. It raises important issues about privacy and freedom that should certainly be thought about. However, the more pressing issue is one of how we have come to be such a camera-surveilled nation. If we understand the history of why we have become so camera-surveilled we can start to question whether the amount of CCTV we have is necessary, whether it is useful and effective, or whether there are other less intrusive options, and potentially more effective, options available. CCTV was not always a surveillance technology. It has come to be synonymous with providing security and safety. However, at its inception it was a multi-purpose technology. Its surveillance capacity was constructed through its applications, with a variety of political, social, private, and economic forces constructing and defining its development and subsequent uses.

This chapter develops a history of CCTV in the UK from the 1950s. Currently, there is little research into the history of CCTV in the UK, covering the economic, social, and political factors involved in its dissemination. Some authors have suggested looking back to the 1990s to explain the widespread use of CCTV (see for example, Goold 2004). However, there is a need to go back further. Williams (2003), for example, looks at police use of CCTV from the 1950s. I attempt to add to this research and believe strongly that looking at the longer history of the use of CCTV can increase our understanding of the reasons why the technology has become so widespread. In this section, I take a step back, tracing the history of CCTV from the 1950s, in order to chart the development of the technology from what I will show to have been a tool for the education, transport and medical sectors, to one of crime prevention and control, and eventually politics.

A History of CCTV – Early Uses

The first mention of the technology I have found is contained in an article in *The Times*, which describes the transmission of an inauguration ceremony of a British Oxygen Company factory, through a closed-circuit television link between London and South Wales in 1956 (*The Times* 2 August 1956). In November of the same year, CCTV was mentioned as relaying a ceremony, with Princess Margaret in attendance, at North Staffordshire University College, into the buildings adjoining the conference hall (*The Times* 21 November 1956). CCTV at this time was used as a technology to transmit moving images from one place to another. This use of closed-circuit television as a point-to-point transmission technology also occured numerous times during 1956 and 1957, in a variety of contexts such as race courses, boat clubs and auction houses, thereby expanding the area within which spectators could witness significant events in real time (*The Times* 31 November 1956; *The Times* 31 February 1957; *The Times* 11 July 1957).

In January 1958, closed-circuit television was used by the Labour Party to test how they appeared on screen and to 'improve the quality of their political broadcasts', therefore using CCTV for training purposes (*The Times* 23 January 1958). The same development is reported by the *Daily Mirror*; however, they also described CCTV being used by the Conservative Party for the last eighteen months. They described CCTV used in this context as a 'vital force in politics' adding that 'five minutes on TV is worth more to the politicians than five years of speeches in draughty halls' (*Daily Mirror* 23 January 1958). Throughout 1958, CCTV is reported as increasingly being used in the business and finance industries, to transmit information (for example, from the Stock Exchange) to television receivers in offices (*The Times* 16 May 1957; *The Times* 3 November 1958). The technology was reported as being very expensive and that it was for this reason that it was not put to greater use throughout the City of London at this time. CCTV was also used during the same year to transmit on screen coach departures and arrivals to an air terminal in London (*The Times* 26 May 1958). A few months later, in August 1958, closed-circuit television was installed in a church in order to transmit the service to another hall, for any overflow of congregation (*The Times* 12 August 1958).

In November 1958, the *Daily Mirror* reported that CCTV had been used in the United States to detect shoplifters and that there was potential for the technology to be used in the same way in Britain. However, the article reported this development as a failure as: 'Customers crowd round the cameras and wave wildly, thinking they are being televised by a national network. Result: Chaos in the store and a lot of indignant shoppers when they discover that no giveaway prizes are forthcoming. Management are now planning to hide the cameras' (*Daily Mirror* 16 November 1958). It is interesting that the US started to use CCTV for the purpose of catching shoplifters as early as 1958, when it seems the idea had not yet reached Britain (yet the trajectory of the use and diffusion of CCTV then became very different). The same year, an article reported CCTV being used to tackle vandalism at Elm Park Station in Essex. The article described the installation of a CCTV camera on the platform, which could be 'swung and tilted automatically up and down the line and a 'zoom' lens can give enlarged views of suspicious movements on the track' (and the camera was reported as also being linked to a screen and controls in the staff room) (*The Times* 25 June 1965).

In March 1959, a closed-circuit television system was installed at King's Cross train station in order to transmit messages from the signal box to the public (by a member of staff receiving the transmission of information from the signal box, writing the information received on a board in front of the transmitting unit). This development was cited as: 'The first time closed-circuit television has been used for this purpose' (*The Times* 6 March 1959). Another article stated that: 'Something similar is already used for relaying information on an American Air Force station in Britain, at a stockbroker's office and on a South African racecourse, but British Railways are the first railway undertaking to use it' (*The Times* 12 March 1959).

A number of articles during 1958 and 1959 reported CCTV being used for educational purposes, primarily in the medical sector. One article detailed an operation, which was filmed and transmitted to a medical conference in Liverpool (and in colour, which was a new development) (*The Times* 30 October 1958). Another article reported the installation of a permanent closed-circuit system in Birmingham University Faculty of Medicine, to transmit images from an operating theatre to a teaching hospital, and to a lecture theatre in the university, to allow students to watch surgery demonstrations (*The Times* 3 December 1959). The educational benefits of using CCTV in this way were praised in another article as allowing 'more realistic teaching of large groups of students' (*The Times* 16 October 1959). In these examples it can be seen that there was a mass aspect to the appeal and driving force of CCTV, even in education.

Closed-circuit television was also used from 1956 for the purpose of improving safety in a number of workplaces and for a number of jobs. Reports suggested that it was used by the nuclear industry to look for any defects within reactors (*The Times* 17 October 1956). Another article mentioned a 'mechanical eye' used to aid in the construction of buildings (to allow the assessment of the state of the earth within holes drilled without physically sending an engineer down) (*The Times* 3 October 1957).

In 1959 an article appeared detailing the intention to install closed-circuit television onto public transport vehicles as a 'safety device'. The technology was described as an 'extra eye for the driver'. Although this article referred to the enhancement of passenger safety, the connotation of surveillance was also apparent, due to the 'extra eye' allowing the driver to surveille the passengers (*The Times* 18 June 1957).

In August 1965, *The Times* published an article detailing the planned installation of two CCTV cameras by Camden Borough Council in tall blocks of flats so that 'mothers can keep watch on their children in the playgrounds below', with parents able to view footage from the playgrounds through a standard receiver and two extra channels on their television (*The Times* 21 August 1965). A year later the project was reported as having failed (with a scathing indictment by the journalist of wasting money on the experiment) (*The Times* 25 August 1966).

In 1967, the company Photoscan was the first company to introduce CCTV to British retailers. The British Security Industry Association, a trade association for the professional security industry, was also established in 1969 (Moran 1998, p.279). In January 1970, *The Times* published an article describing a visit to Britain from the Photoscan Chairman – R.C. Hibbard – to 'explain some of the applications of videotape recording to crime prevention in the US'. The Photoscan 'eye' was described as a 'multi-lensed black hemisphere, which is suspended from the ceilings of stores, shops and supermarkets and transmits pictures by closed-circuit television to a monitor screen in the security manager's office'. This type of CCTV camera was in the shape of a black ball – only pointing in one direction at once. However, members of the public could not see which way the camera was pointing (the shape would suggest that it had a circular view, rather than being fixed on one area/point at a time).

CCTV began to be installed in the mid-1970s in various London Underground stations, including: Clapham Common, Stockwell, Clapham North and Brixton underground stations (Mayhew 1979, pp.21–3). The reasons given for the installation of these systems were to deal with assaults on staff and to combat theft in underground stations (Webb and Laycock 1992, p.11). The 1970s also saw the introduction of videocassette recorders using analogue technology, with the capacity to record and store recordings on tape. This technology was not very effective in low light or at night-time, however there were improvements in terms of the general reliability of the technology (Fennelly 2003, p.210).

Charged Coupled Device cameras were brought into use during the 1980s, allowing low light and night-time recordings. The introduction of these solid-state CCTV cameras was a significant development, replacing most of the tube type cameras by the early 1990s. Advances in Digital Multiplexing allowed recordings on multiple cameras, as well as new capabilities such as time-lapse. More recently (and since the late 1990s), digital technology has lowered the cost of cameras (although they are still considered to be very expensive by some), in addition to providing greater clarity and extra capabilities such as zoom (Roberts 2004).

The Data Protection Act came into force during 1984, the same year that cameras were used for the purpose of crowd control during the miner's strike. This was not a permanent CCTV system installation, yet as I have argued in the previous chapter, I believe that it signifies the beginning of a more widespread use of video surveillance. It is tempting to use pivotal moments when writing a history of something, and I don't want to say that it was at 'this moment' that CCTV escalated but I do believe that its use in terms of this political situation signifies the start of something greater in terms of the political construction of the technology.

CCTV in Britain from the 1980s

This section looks at the use of CCTV from the 1980s. It focuses on the use of CCTV for a variety of purposes, and includes the use of CCTV by the police from the 1980s onwards (picking up where the previous chapter left off with the 1984 Miners' Strike).

CCTV was installed in the early 1980s at the Dartford Tunnel. Initially used for the purpose of catching stolen vehicles, it was also used to monitor the movements of coal miners on strike travelling north from Kent (Emmerson 2000, p.523). Using cameras to monitor people continued and expanded during the 1980s. With regard to popular culture, the television programme *Crimewatch* was aired for the first time in June 1984. The programme featured surveillance footage from CCTV cameras and photographs of suspects to the public (*The Independent* 21 April 2000). With reference to *Crimewatch*, Norris and Armstrong (1999, p.69) argue that:

> CCTV works because it deters crime, and if it does not deter crime it enables police to be deployed and apprehend a suspect, and if it does not result in immediate apprehension, in the absence of the police knowing the identity of the culprit, then the public can provide it. And the proof is there before the viewers' very own eyes. For all the scripted words and statements in praise of CCTV, it is perhaps the continuous repetition of the visual narrative structure of success which makes such programmes so seductively powerful in promoting CCTV.

Davies (1996, p.186) suggests that a year later, CCTV was installed in football grounds, which he argues is the beginning of the CCTV trend. Grants were allocated by the Football Trust to install these systems, for example Sheffield United was given £25,000 to install 15 cameras (Armstrong 2003, p.117). Alongside this, the police were also allocated funds to set up mobile CCTV units. In 1989 the Football Trust provided South Yorkshire Police with a grant of £30,000 for mobile CCTV cameras. Davies argues that a lack of legislative guidelines restricting how and where the cameras could be installed meant that the police found more uses for the systems. He goes on to suggest that word of mouth played a major part in CCTV being used by an increasing number of people involved in law enforcement, and that it was this initial promotion and support for CCTV that explains why Britain has so much video surveillance.

McGrath (2004) also argues that it is the links between football hooliganism and CCTV that allowed the technology to really 'take off'. He suggests that:

> The history of the surveillance camera in the UK is inextricably tied up with the football hooligan and the various cultural responses to that phenomenon. Surveillance cameras were first introduced to grounds in the mid-1980s as part of a Thatcherite attempt to curb a perceived crisis in football-related violence ... By the end of the 1980s, crowd surveillance at football matches was routine.

Although this may partly explain the beginning of the establishment of CCTV in Britain, there are other reasons for its rapid uptake and more widespread dissemination in terms of its early history. We have seen various attempts by industry to sell cameras for a range of purposes, and to tackle a range of issues. Police take up of the technology was sporadic prior to this time, but CCTV was starting to be used for crime prevention, as well as for other purposes such as traffic control. Manwaring-White (1983, p.91) also describes the use of video surveillance by the police at football matches as far back as 1977. Although 1985 may be the date of permanent installation into football grounds and a rise in funding for CCTV for this purpose, it was being used for the same purposes prior to this date. I also found one other instance of the use of CCTV in football grounds even earlier than Manwaring-White suggests, in *The Times* archive. As early as November 1968, the newspaper included an article on the installation of CCTV to combat hooliganism:

> Closed circuit television is now working on some grounds: the police are able to scrutinize the crowd as they flood through the turnstiles. (*The Times* 26 November 1968)

In 1985, a CCTV system for monitoring public space was installed in Bournemouth (Moran 1998, p.280). This development is noted by a number of commentators as being the first public space system in Britain. However, as I discussed in the previous chapter, there are earlier examples, such as Liverpool in 1964/1965 and Trafalgar Square in 1968. Some suggest that the reasons given for the installation of this system was the Conservative Party conference. At the party conference in 1984, an IRA bomb targeting Margaret Thatcher had exploded and killed 5 people (the then Prime Minister was unhurt). This installation of cameras, they argue, was a response to the IRA threat (Norris et al. 2004, pp.110–35). Others argue instead that local council in Bournemouth had concerns over street safety and that this is the reason the system was installed (Ainley 2001, p.86).

In January 1986, Margaret Thatcher held a seminar on crime prevention, which resulted in the establishment of a working group to conduct a study of how to reduce crime on the London Underground. One of the main recommendations was for further and improved installation of CCTV in stations (Webb and Laycock 1992). In 1987, a CCTV scheme began in King's Lynn in Norfolk, set up on an industrial estate, with the rationale for installing the CCTV system described as an attempt to control burglary rates. The second stage of the scheme in King's Lynn was the introduction of 19 CCTV cameras to various sites in the town centre, including car parks and access to buildings (Brown 1995, p.60). Five years later, 32 cameras were set up in car parks, and the scheme was extended out to housing estates and the city centre (Davies 1996, p.176). Regarding this development, Davies argues that the rationale behind it was not readily apparent, due to King's Lynn having a relatively low crime rate (particularly in terms of a national comparison) and quotes the surveillance project director as saying:

> What it comes down to is a perception of crime, rather than the crime itself ... the surveillance system has grown because of the 'feel-good' factor it creates among the public. (ibid.)

In a twist of Bentham's notion that it no longer matters whether you are actually being watched or not, people simply fear that they are and adjust their behaviour accordingly. The quote above shows a popular political representation of CCTV – it doesn't matter whether the technology has any impact on crime rates and actually makes people safer, it is simply enough to make them *feel* that they are safer. I come back to this in my next chapter.

The Safer Cities Programme was launched in 1988. Concurrently, the organisation 'Crime Prevention' was also established (Goold 2004, p.32). In comparison with the rest of Europe, the introduction of CCTV and its subsequent growth during the 1980s and early 1990s in Britain met with little opposition. During this time there was a more general rise in levels of surveillance, including peer-to-peer surveillance. For example, the launch of 'Crime Stoppers' in 1989, which encouraged people to anonymously report any details of crimes that they had information about. At the same time, the Department of Social Security (as it was known then) launched their benefit fraud hotline, which enabled people to anonymously report those they believed to be involved in benefit fraud (Norris and Armstrong 1999, p.30).

Goold (2004, pp.19–20) argues that during the early 1990s the police in the UK were not heavily associated with CCTV systems for public space preferring to wait for the public reaction before becoming involved. He says that the police were 'keen to avoid being portrayed as a "Big Brother" figure ... such fears eventually led the Association of Chief Police Officers (ACPO) to advise Chief Constables in 1993 to take a "back-seat" role in the promotion of CCTV and to hold off committing resources to the technology until the police were better able to gauge the public reaction to the introduction of cameras'. He follows this by saying that as CCTV rose in popularity in terms of politicians and the media; police forces became more active in their promotion and installation of systems, culminating, in the mid-1990s with police as 'leaders or active partners in CCTV schemes all over the country'. Fay argues that the primary reason for the widespread use of CCTV has been its 'promotion as a panacea for a wide range of social and economic problems by a variety of state agencies and commercial organizations, often working in partnership' (1998, p.316). This is reiterated by Goold (2004, p.20) who argues that 'public area CCTV [emerged] at a time when politicians and policy-makers were in search of a new solution to the problem of crime and, perhaps more importantly, a way of convincing the public that they were serious about crime prevention'. Although, as we have seen in the previous chapter, there were a number of reasons why police take up of the technology was hindered prior to the 1990s (cost, differing objectives from the Home Office and the police, and technical difficulties).

The first speed cameras in the UK were installed in West London in 1992. In the same year, a Home Office report on the effectiveness and acceptability of CCTV was published (Honess and Charman 1992). The report depicted a public generally unconcerned with the appearance of CCTV in public spaces. However, it also showed that fewer than half the people surveyed believed they would feel safer following the installation of CCTV onto public streets, car parks and shopping centres.

The murder of James Bulger in 1993 was an important incident in relation to the history of CCTV and public perceptions of the technology (although there are other examples of the use of CCTV in relation to high profile child abduction cases, which I cover in Chapter 5). James Bulger (a two-year-old) was abducted in a Liverpool shopping centre in 1993 by Robert Thompson and Jon Venables (both ten years old at the time). His body was found two days later on a railway line in Bootle (a town a few miles away from Liverpool city centre). Robert Thompson and Jon Venables were charged with his murder. The CCTV images of the killers leading James away were shown on the news, and media coverage repeatedly stated that the killers had been captured on CCTV. Although CCTV did not stop the incident occurring (it aided in identification), following this time it was extremely difficult for anyone to oppose the technology and those that did were seen as being on the side of the killers (see Davies 1996; Norris and Armstrong 1999).

McGrath also discusses this incident and the use of CCTV. He explains that at the time of being shown the CCTV footage for the first time (after the incident and before the body had been found), Denise Bulger thought that it meant that her son was safe and was reassured by it (Jon Venables was holding James's hand and therefore could be perceived as looking caring). McGrath goes on to suggest that: 'The violent act of kidnapping, much loved by Hollywood for its dramatic bursts of action, here does not look violent at all, and hence is not noticed. This unreadability of the image contributed to a public sense of trauma, of helplessness, in relation to the case' (2004, p.35). This is an interesting take on the idea of CCTV. If Denise Bulger was reassured by the images shown to her, were the public also? Did this go further than simply not being able to argue against the use of CCTV for fear of being seen to be on the side of the abductors, but instead moved towards cementing an idea and ideal of safety and security (provided by CCTV) in the minds of the public? Resting on the notion of a protective all-seeing eye that is somehow reassuring (even though it doesn't stop the incident happening), as the public never see the next part – the story for them ends there (at least visually).

In the aftermath of the murder of James Bulger, John Major (then British Prime Minister) announced in an interview with the *Mail on Sunday* that: 'Society needs to condemn a little more and understand a little less', and: 'My concern is to be considerate to the victim and protect the victim rather than be considerate to the criminal and forgive the criminal' (*Mail on Sunday* 21 February 1993). It could be argued that the murder of James Bulger started to become politically symbolic during this time. In terms of political discourse, the James Bulger murder was described as being indicative of greater societal problems (I will also come back to this point in Chapter 5 in terms of the media and moral panics). In a 1993 *Sun* article, Tony Blair (Shadow Home Secretary at the time) wrote:

> It is not simply the horrific nature of the crimes themselves that have shocked Britons in these past few weeks. It is the feeling, difficult to define, but powerfully with us, that they are symbolic of a deep malaise in our country. (*The Sun* 3 March 1993)

It could be suggested that the constant replaying of CCTV images not only imprinted a usefulness on the technology in terms of the James Bulger case, but that it also began to be tied up with a wider political discourse at the time of a decline in social morality and the problem of rising crime.

The IRA detonated two bombs in the City of London, one in April 1992 and the other in April 1993. The first resulted in one casualty and £350 million worth of damage (*International Herald Tribune* 1998). The second (in Bishopsgate) also resulted in one casualty, 40 injuries and £650 million worth of damage (*BBC News* 8 July 2000). As a result of these incidents more CCTV systems were installed throughout the City of London, roadblocks were put in place, and there was an increase in visible policing and increased security at building entry points. This became known as the 'Ring of Steel' (Coaffee 2004, p.278). During the development of the 'Ring of Steel', the Chief Constable for City of London, Owen Kelly stated: 'If lives are to be saved by the use of these cameras, I cannot see that there can be any objection to them'. The Chairman of the Association of Chief Police Officers' committee on terrorism concurred, and stated: 'There will be an increased use of video cameras in the future and what we are finding is that the pressure for them is coming from the local authorities'. The head of the Home Office Crime Prevention Unit argued: 'You hear emotional pleas about Big Brother watching us but, in fact, there are very few objections from the public' (*The Guardian* 13 May 1993).

In September 1993, 'Camera Watch' – a scheme developed by the police to encourage more private CCTV systems to be set up in the City – was established, and by the end of the year CCTV systems were running in 39 UK towns and cities (Coaffee 2004, p.282). Earlier in the year, on the back of Exeter and Birmingham City Councils disallowing two applications for CCTV systems, Downing Street announced that the installation of camera systems would be exempt from requiring planning permission, which in turn had the effect of taking the authority away from local councils in terms of whether they wanted systems installed or not (Davies 1996, p.187). Of this development, one government minister argued that it would get rid of any 'unnecessary red tape' (Fay 1998, p.317).

In 1994, the Home Office published its report, 'CCTV: Looking out for You' (which is looked at in more detail in the next chapter). On 5 July, as a 20 camera CCTV system was launched in Liverpool city centre, the junior Home Office Minister David Mclean announced:

> This is a friendly eye in the sky. There is nothing sinister about it and the innocent have nothing to fear. It will put criminals on the run and the evidence will be clear to see. (*The Times* 6 July 1994)

In October of the same year, the Home Office announced £2 million worth of funding for CCTV by way of a Challenge Competition. By the end of the year, the number of towns and cities with CCTV systems had risen to 79. By March 1995, this number had increased to 90 (Goold 2004, p.17). On 22 November 1995, Home Secretary Michael Howard launched a £15 million CCTV scheme competition, announcing at the same time the details of the Crime Prevention Agency set up to 'reduce the fear of crime' (*The Guardian* 23 November 1995). This agency comprised the Home Office, the police and Crime Concern (a national crime reduction organisation). The competition was to 'encourage the expansion of CCTV', with 800 bids received. There had also already been a £5 million project six months earlier (Jones and Newburn 1998, p.62). CCTV was also installed for the first time in a residential area (in Newcastle) during the same year (Moran 1998, p.283).

A further bomb in the London Docklands in February 1996 meant that security was immediately increased. Coaffee (2004, p.285) describes the situation from August 1994 (and the Provisional IRA ceasefire) to February 1996 as one of lessening security and even a call from one private business to disband the 'ring of steel' completely. Nevertheless, cameras and CCTV systems were continuously updated and improved, including a CCTV system at entry points into the City with number plate recognition capability, even with the declining IRA threat. Camera technology became 'perhaps the single most important factor in the City of London Police's counter-terrorist campaign' (Coaffee 2004, p286).

In March 1996, the Home Office published the White Paper, 'Protecting the Public'. The paper included the statement:

> New technology is moving to the forefront of the fight against crime. Closed circuit television (CCTV) surveillance systems have proved very effective in preventing and detecting crime and in deterring criminals; CCTV is overwhelmingly popular with the operators, the police and the general public; Clearly CCTV works. (Home Office 1996)

It also promised £45 million additional funding for CCTV schemes and to provide 10,000 more cameras over the next three years. In terms of the Government's broader approach to crime prevention, the White Paper emphasised a co-ordinated approach to reducing crime and a local and community partnership approach (including Neighbourhood Watch). It also stated that it would 'wage war on crime' (p.54). In the same year, the Scottish Office released their White Paper entitled 'Crime and Punishment' in which they wrote:

> There is clear evidence that ... [CCTV] cuts crime ... The Government will foster the use of CCTV in many more communities in Scotland. (The Scottish Office 1996, pp.13–14)

The CCTV Challenge Competition round one results were also announced in 1996, by which time it has been estimated there were over 200 schemes in the UK. Following this, round two of the competition was for funding for the years 1996–97, round 3 for 1997–98, and round four for 1998–99 (Hansard 18 June 2008).

In October 1998, the first CCTV system with facial recognition capability was trialled in the London Borough of Newham (*The Observer* 11 October 1998). This system used 140 cameras to scan the faces of pedestrians. These images were run through a database in an attempt to match them to known criminals (Brin 1998, p.5). Algorithmic technology, carried out in real-time, was used to detect and identify human faces from CCTV footage (Emmerson 2000, p.522). However, facial recognition technology at this stage was not hugely effective.

In August 2000, the Home Office published the results of an evaluation showing street lighting to be up to seven times more effective than CCTV in reducing crime (Welsh and Farrington 2002). However, only a year later, the Home Office announced £108 million worth of funding for 300 new CCTV schemes (*The Independent* 31 March 2001). The following year, a £3 million CCTV control centre (controlling 400 cameras) was set up in Manchester's city centre (*The Independent* 29 June 2002).

In November 2001, two months after the terrorist attacks in the United States, the Anti-terrorism, Crime and Security Act was formally introduced in the UK.[1] The Act allowed government departments to collect and share information on terrorist activities, and enhanced police powers to deal with those in custody unwilling to share their identity (Davies 2005, p.xxviii). The 2001 Act was replaced by the 2005 Prevention of Terrorism Act, allowing the Home Secretary the right to apply a control order (with the effect of restricting an individual's liberty) on any individual suspected to be involved with terrorism. In 2003, the 'Ring of Steel' was extended into Holborn and Victoria Embankment (*BBC News* 18 December 2003).

The Home Office published a report in 2005 outlining the results of evaluations of 14 CCTV systems. The findings showed that CCTV does not significantly reduce crime in public areas (Gill et al. 2005). On the 20 September 2005, Scotland Yard announced that CCTV filmed the 7 July bombers carrying out a 'dry run' on the London Underground just over a week before the attacks, essentially showing that CCTV in this case was ineffective in preventing a terrorist attack. However, this was not the only way this footage could have been interpreted. For some, it could once again (as was the case with James Bulger) cement the usefulness of

1 The terrorist attacks in New York took place on 11 September 2001. They were a series of co-ordinated suicide attacks by al-Qaeda (a Sunni Islamist terrorist group) on a number of buildings in the United States.

CCTV as a way of capturing images of the bombers, and providing evidence for identification. In October 2007, the Home Office and Association of Chief Police Officers jointly published their National CCTV Strategy. The report stated that:

> Closed circuit television plays a significant role in protecting the public and assisting the police in the investigation of crime … CCTV in the UK enjoys significant public support. (Gerrard et al. 2007)

The report goes on, however, to suggest that CCTV had not been used to its full potential to date due to being developed and operated in a 'piecemeal fashion', hence the need for a national strategy. The report also suggested that the location of cameras could be improved and that picture quality varied and was often poor. It is interesting to compare the assessment of CCTV revealed in this report, with the negative aspects highlighted with regard to the 1996 Home Office report, which stated that 'CCTV works'. This claim had clearly been weakened by the actual results produced, the purposes for which CCTV had been used, and how effective it had been in cutting crime. I come back to this report in more detail in the next chapter.

By April 2007, new 'talking CCTV cameras' had been set up in Middlesbrough town centre (trialled in September 2006), with plans to install the systems in Southwark, Barking and Dagenham, in London, Reading, Harlow, Norwich, Ipswich, Plymouth, Gloucester, Derby, Northampton, Mansfield, Nottingham, Coventry, Sandwell, Wirral, Blackpool, Salford, South Tyneside and Darlington. The Middlesbrough system consisted of 12 'talking cameras', operated by staff in a control room. The primary aim of these cameras was to stop littering and anti-social behaviour (*BBC News* 4 April 2007). Barry Coppinger, Middlesbrough Council's executive member for community safety, was reported at the time as rationalising these systems:

> CCTV is making Middlesbrough safer and this added facility will make it more effective. For example, if an operative now sees someone dropping litter, they can tell them to pick it up, or if they see an incident starting to get out of hand, they can give advice that will hopefully nip it in the bud. I think that it will give people extra confidence as they go about their business and re-enforce the message that Middlesbrough is a place that is constantly thinking about community safety. (*BBC News* 17 September 2006)

The systems sparked a strong reaction from civil liberties campaigners, with Liberty describing them as a 'waste of money' (NoCCTV 29 July 2011). A newer system designed in 2012 to be installed in residential areas has been described as belonging only in a 'police state'. Nick Pickles of Big Brother Watch stated that in terms of the new systems being installed in Camden, London:

> This kind of technology may be acceptable in a police state or a science fiction film, but it is absolutely not in modern Britain ... Proper regulation of how and when this kind of equipment is installed is badly needed, and it should simply not be used in residential areas ... The idea that a Robocop recording will tackle antisocial behaviour and crime is as laughable as it is a total invasion of privacy. Who knew councils had the authority to take your photograph simply because you walked into a communal garden? (*The Telegraph* 6 February 2012)

The 'talking' part of these cameras is now being switched off. This has also happened in other locations across the country since 2011 (ibid.).

In 2008, DCI Mick Neville (Visual Images, Identifications and Detections Office) suggested that only 3% of street robberies in London are solved through the use of CCTV. The VIIDO was piloting a new database of CCTV images at the time (*The Guardian* 6 May 2008). A year later, Neville estimated that £500 million has been spent on CCTV by the government, with over 1 million cameras in London and 1 crime in 1000 solved per year by CCTV (*The Telegraph* 20 August 2013). In May 2012, a Freedom of Information request by Big Brother Watch revealed that £4.1 million had been spent on VIIDO in the past year (Big Brother Watch 2012).

However, since making these statements, Mick Neville has also talked about the use of 'super recognisers' who watch footage from CCTV cameras and pick out 'known' offenders. Since 2011, over 200 police officers have been recruited as 'super recognisers', due to their ability to identify criminals based on their images. This initiative has been dismissed as the 'latest gimmick' to justify the widespread use of video surveillance by civil liberties groups (*Daily Mail* 27 September 2013).

This year (2013), Minister for Criminal Information, Lord Taylor of Holbeach, stated that CCTV (and ANPR) are 'vital tools' when discussing the new CCTV Code of Conduct (*BBC News* 12 August 2013).

The last few years have seen a range of technical capabilities increase in terms of CCTV systems, with the inclusion of a variety of audio and video analytics, and network cameras with HDTV performance.

What Does this History Tell Us?

During the 1950s CCTV was characterised and defined as a transmission technology. It was used in a very literal sense to transmit images from one place to another, via a closed circuit. It was used for people to be able to witness events and images in real-time without having to be in the actual place at the time. In this context, CCTV was constructed as being another form of TV. This period also saw CCTV being used to improve political broadcasts – used by politicians to practice speeches and to see how these would look on television. It was used for educational purposes in the medical sector, to allow students to view operations in real-time. Once again, we see CCTV being used as a form of television. During this period CCTV was also used for health and safety reasons. The discourse surrounding the technology was that it was a 'mechanical eye'. This discourse was echoed with regard to the use of CCTV on public transport. Described as an 'extra eye' for the driver, the technology was defined as something mundane, ordinary and personified (it was given a human trait).

The 1960s are characterised by various attempts to use CCTV for crime prevention purposes (although as we have seen in the previous chapter, these attempts were hindered for reasons of cost and efficiency). Experiments to target criminal activity in 'crime hot spots' began to see CCTV being used, in theory at least, as a surveillance technology (although has been shown previously, these experiments proved to have little worth over a policeman sitting in a building with a pair of binoculars). Aside from uses by the police there was also other evidence of attempts to move CCTV from a transmission technology to one used for the purposes of surveillance. The proposal to install a CCTV system into a playground in Camden echoed the discourse found in the previous period. The technology was there to enable parents to keep a 'watchful eye' over their children. In this sense, CCTV was once again situated and constructed as a mundane technology, simply there to provide an extra eye.

During the 1970s police use of the technology remained limited. It was most often used for traffic control purposes but remained too costly to be put into general use. The technology did start to be used for crime prevention purposes during this time in contained spaces, with its installation into a number of London Underground stations. There is also evidence that it was beginning to be used to combat football hooliganism, with the introduction of CCTV systems into football grounds and CCTV being used to enable police officers to watch the crowds entering the grounds.

The 1980s are a period characterised by a move towards CCTV becoming situated more firmly as a surveillance technology. Although we saw a potential move towards this type of use during the 1960s, various elements stood in the way and hindered its development. During the 1980s the use of the technology for surveillance purposes no longer remained a possibility but started to be utilised more permanently for reasons of crime prevention and detection (at least in theory) and for political purposes. As I discussed in the previous chapter, CCTV was installed in residential areas during the 1984 Miners' Strike to surveille the movements of miners (and in turn also the other residents). A year later it was installed into a number of football grounds to combat hooliganism. It was also during this time that *Crimewatch* started to use footage from CCTV cameras – showing images to members of the public of crimes committed and captured on CCTV. Members of the public started to become involved in a network of television, the police, and CCTV to catch criminals. Later pinnacle moments (such as high profile child abduction, or terrorist incidents, which I discuss in Chapter 5) have been captured on CCTV and replayed to a public audience, and have thereby contributed to the idea of CCTV as a useful technology to combat criminal activity. Through the use of the media the technology has become constructed as a tool for evidence. Despite its main purpose continuously cited as 'crime prevention', it becomes defined, through its use as an evidentiary tool, as useful and effective whether or not it achieves its primary aim.

The 1980s also witnessed the start of the trend to use CCTV to make people feel safer. The example of the installation of a video surveillance system into King's Lynn (an area with relatively low crime rates) shows the emergence of a discourse that still runs strongly through present day Home Office reports (as well as in other policy documentation) – that CCTV makes people feel safer, and that it promotes a 'feel good' factor in members of the public. It was during this time that we see CCTV being defined as a technology to provide feelings of safety. Whether or not it produced results in relation to its primary purpose of crime prevention, it fulfils its obligations under a new justification of providing something of benefit to members of the public.

During the 1990s CCTV was increasingly installed in public space. This period is also characterised by its links with high profile criminal investigations (I mentioned James Bulger in this chapter, but return to this point with further examples in Chapter 5). It was also a time when the language used by the Home Office and the police was extremely positive in terms of the uses, capabilities, and effectiveness of CCTV. During this period, CCTV is defined as a technology to provide safety and security, and civil liberties concerns were readily dismissed. Grand claims were made during this time regarding CCTV as providing a deterrent to criminal activity, and suggestions that 'the evidence will be clear to see'. Discourse surrounding CCTV concentrated on the image of CCTV as something that 'works', and that it had proven to be 'very effective in preventing and detecting crime'. The dominant persuasive discourse (Pfaffenberger 1992) therefore became one of protection, safety and security, and effectiveness.

During the 2000s (and to the present day) evidence started to arrive that the grand claims made by the Home Office and the police regarding the effectiveness of CCTV did not necessarily hold up under scrutiny. In terms of the claims of effectiveness in cutting crime, evidence arrived stating that street lighting is seven times more effective at reducing crime. Other reports suggested that CCTV does not significantly reduce crime in public areas. The 2007 National Strategy directly contradicted claims that 'the evidence would be clear to see', stating that picture quality from CCTV footage was varied and often poor. There are a number of other studies (including Home Office funded studies) that provided evidence that goes against the discourse and rhetoric (concentrating on the effectiveness of CCTV) found in the 1990s and up to the present day. Despite this evidence, CCTV continues to be cited as a technology that 'works'. I will come back to this again in the next chapter.

Conclusion

In this chapter I have outlined a history of CCTV in terms of the social, economic, political and technological developments contributing to its introduction and subsequent widespread use in the UK. I have charted the development of CCTV from a technology used for education, transport and medical sector purposes (which I mentioned was a neglected area of research), to a tool for surveillance, crime prevention and criminal justice. I have shown in this chapter and the previous one, that CCTV was described as potentially being useful for crime prevention purposes, to the police by industry, as early as 1958. During this same year, CCTV was installed as an 'extra eye' on public transport vehicles. Its use as a surveillance method is shown again a few years later, in 1965 in a block of flats for reasons of surveilling the playground. During the 1960s, CCTV was used by the police in public areas, although technical difficulties hindered the trials. I have shown through the ACPO archives that there is a real 'push' from industry for an uptake of the technology. However, the police at this time were less willing. Financial constraints certainly played a part in this, although technical issues were also apparent at the time. I have described the technological developments of the 1970s, occurring alongside the introduction of cameras onto London Underground. The 1980s saw an increase in public space monitoring and the start of its use as a preventative and exclusionary technology; in the context of the miners' strike and at football grounds. I have shown that the 1990s saw a financial commitment from the government for increased CCTV, occurring alongside incidents such as the James Bulger murder, and a rise in fear of crime, which meant that the environment was ripe for increased use of the technology; if only to persuade the public that the government was tackling (perceived) rising levels of crime. More recently, the rationale for the use of CCTV has centred on preventing crime and terrorism, with continued financial commitment from the UK Government for the technology.

In terms of key events the James Bulger incident was clearly a factor contributing to the rise of CCTV; a moral panic coupled with rising crime rates and increasing fear of crime by the public were certainly factors prompting an increase in budget for CCTV cameras. However, this was linked to a general redirection of criminal justice policy – CCTV, like any technology, does not develop in isolation. I showed in the criminal justice section that both the Conservative and Labour parties adopted a hard line stance during the 1990s, with regard to crime and crime prevention. CCTV as a technology fitted easily into this promise of tackling crime, in terms of enabling a technological solution to a wider societal problem. At least on the surface, both parties could announce that something was being done.

Chapter 5
Political and Media Discourse

Introduction

> [CCTV] is a friendly eye in the sky. (David Maclean, Home Office Minister, 1994)

78 per cent of the crime prevention budget was spent on CCTV during the 1990s. A further £500 million was spent between 2000 and 2006 (House of Lords 2009). The total cost of installing and operating CCTV systems incurred by Local Authorities was estimated at £515 million between 2007 and 2011 (Big Brother Watch 2012). We are told that the 'innocent' have nothing to fear from CCTV cameras. The idea that CCTV and other surveillance technologies could potentially infringe on our civil liberties is dismissed in public discourse with a counter-claim that these technologies benefit us in other ways (for example, they provide protection from terrorists), and therefore we are encouraged to believe that there is nothing to discuss, or to have concerns about. Surveillance equals security, and people should simply be willing to give up privacy in order to obtain security. It is portrayed as a zero-sum game. Giving up our privacy, we are told, equals an immeasurable gain in security and safety.

In this chapter, I look at the political and media discourse surrounding CCTV. I include a review of the academic and Home Office literature focusing on public attitudes, awareness and acceptance of video surveillance systems, in a variety of contexts. I also focus on public engagement in relation to CCTV, and argue that the public have been sidelined in any consultation or strategy development. A strong political and media discourse centring on ideas of safety and security has seemingly overridden any concerns the public might have about a loss of privacy due to surveillance systems. Perhaps even more importantly, there is a lack of debate concerning the amount of public money spent on systems that have been found to fall short in achieving the aims for which they are installed. When debates in the area of CCTV do arise, they are often centred on two sides – one arguing that CCTV provides safety and questioning why anyone would be against that – and the other arguing that CCTV infringes on privacy and civil liberties. There is an important point to consider, however. Why are we spending such a large percentage of our crime prevention budget on a technology that has been proven not to work effectively in preventing crime?

Studies of CCTV

Over the last 20 years there have been numerous academic studies published disputing the effectiveness of CCTV. Despite this empirical evidence, public attitudes appear to be consistently positive about the widespread use of CCTV (although it could be argued that Home Office funded studies show bias towards acceptance of CCTV and lack impartiality) and funding for CCTV systems has continued or increased. In fact there are even examples of public pressure on government and local authorities to install more CCTV into residential areas and town centres by residents associations (for a detailed empirical account, see Fussey 2004). At the time of the Home Office CCTV Challenge Competitions (outlined in Chapter 4), partnerships bidding for funding for CCTV came under pressure from local communities to install CCTV systems in their area, despite these partnerships often being unaware of local crime and disorder problems and how CCTV might help to combat them (Gill et al. 2003). These public pressures continued (with a new interest in the potential displacement effects of CCTV) throughout the Government's third major round of funding for CCTV in the form of the Crime Reduction Programme (Home Office), which ran between 1999 and 2003 (with an investment of £170 million into CCTV). During the 1990s the Government allocated and spent almost 80 per cent of its crime prevention budget on implementing CCTV, with a further £500 million being spent on developing and installing CCTV between 2000 and 2006 (House of Lords 2009).

The Effectiveness of CCTV

Researchers on behalf of the Home Office have conducted a number of studies focusing on the effectiveness of CCTV. These are sometimes coupled with surveys of public attitudes towards, and public acceptance of, CCTV systems. Honess and Charman (1992), and Tilley (1993), conducted a couple of early studies under the umbrella of the Home Office Police Research Group. The 1992 report describes the results of a survey, which sought to research public opinion towards CCTV and how those who install and operate them use these systems. The survey asked questions focusing on: public attitudes towards CCTV, public awareness of CCTV systems and any concerns about these systems, responsibility for the installation of systems, the views of the public on access to and use of tapes, and public perceptions of the purpose of CCTV. The report also included crime statistics from four case study areas for the purpose of discussing the effectiveness of CCTV. The conclusions of this study were that CCTV has a 'broadly positive perception from members of the general public. Levels of concern are not high and CCTV is assumed to be effective in crime control' (Honess and Charman 1992). The authors pointed out, however, that these attitudes are held by a public without a sound knowledge of the capabilities and functions of CCTV in public places. They recommend greater provision of information and levels of public consultation, to minimise the potentially negative impact on civil liberties. The survey data is now over 20 years old, as well as having been conducted by non-independent researchers. This call for public consultation has still not been realised. I will return to this point later in the chapter.

A study by the Scottish Office in 1996 also focused on the issue of the effectiveness of CCTV, as well as possible displacement effects. The main findings of this study showed a 21 per cent decrease in the total number of recorded crimes and offences in the area covered by CCTV in the two years following installation (the research was conducted over four years; two prior and two post-installation) and a 16 per cent improvement in the detection of crime. They found no evidence that CCTV had caused any sort of displacement effect (that the crime will happen nonetheless but will simply move to another area not covered by CCTV).

Crime statistics were also the focus of a more recent Home Office funded study outlining the findings from the 2005 national CCTV evaluation. The main areas of focus were: changes in crime, an analysis of other crime reduction initiatives, the design process of CCTV and other factors involved, the installation process; the economic impact of each system, control room operations, and public perceptions of CCTV. The findings from this evaluation pointed to CCTV having little or no effect on crime, as well as being ineffective at making people feel safer or changing their behaviour. The authors argued that to come to these conclusions is too simplistic. They stated that crime rates are a poor measure for assessing the effectiveness of CCTV, because success in reducing one particular type of offence may be swallowed up in the overall statistics. They called for a more balanced judgement of the success of CCTV, proposing that the UK is still learning how to use this technology to its best effectiveness (Gill and Spriggs 2005). The outcomes of this study are not particularly reassuring in terms of the amount of money that has been, and is being, spent on these systems. The proposition that Britain is still working out how to use CCTV is reiterated in other reports, such as the 2007 National CCTV Strategy (seeking to find out how to use CCTV more effectively), which I come back to later in this chapter. It would seem that CCTV remains not only a technology that needs to be used more effectively, but also one whose very use at all can be called into question. The simple fact that the technology is in existence does not justify its use prior to thorough evaluation.

Independent research studies have tended to find lower rates of effectiveness and public acceptance than those funded by the Home Office. For example, Bulos and Sarno conducted the first national evaluation of local authority CCTV initiatives in 1994, with their survey focused on the nature and scope of research into schemes and questioning the extent to which these schemes have extended into the public domain. The evaluation encompassed government studies, an independent evaluation by Liberty, research from CCTV users and providers, and the media. The second part of the report surveyed CCTV use by local authorities, concentrating on: the differing levels of CCTV use, the extent of planned CCTV schemes, and knowledge of research on CCTV to date. The authors concluded that their findings suggested a desire to expand CCTV (and its use in public space) from local authorities and the police, with plans in place to develop and integrate systems. They pointed out that this desire to develop and extend CCTV use comes with no basis in evaluation or research, and therefore called for greater long-term evaluation of CCTV, as well as the establishment of an agency to coordinate activities in public places (p.21). The authors suggested that there are gaps in knowledge regarding the effectiveness of CCTV systems and argued that there has been little independent research into this issue.

A year later, in 1995, a small-scale study was undertaken by Bulos to address the question of whether CCTV has had an impact on levels of crime in Sutton town centre (a suburban high street just south of London), the public perception of CCTV, and how the Sutton scheme compares with others elsewhere. The study found that during the four month period after CCTV installation (June to September 1994), vehicle crime in the three council owned car parks in the town centre fell by 93 per cent, and that there was a 76 per cent reduction in vehicle crime on streets with CCTV. However, the report did provide some evidence of displacement of crime due to CCTV systems.

A more recent Home Office study focused once more on these same questions. Welsh and Farrington (2002) undertook a review of 46 relevant studies of CCTV and a meta-analysis of 22 surveys from the UK and US on the effectiveness of CCTV in reducing crime. The study concluded that CCTV has a minor impact on crime and is most effective in reducing incidents of vehicle-related crime in car parks. They add that it has little or no effect on city centre crime or crime committed on public transport. Included in this paper was a critical analysis of the methodologies used in previous studies, as the authors pointed to what they believed to be necessary future areas of research. They argued that there is a real need for a control condition, coupled with longer-term evaluations with longer follow-up periods, as well as the need for evaluations to determine alternative methods of measuring crime. This study was recently updated, with results showing that CCTV caused a modest decrease in crime (16 per cent) in experimental areas compared to control areas, and the majority of this was due to CCTV in car parks, which resulted in a 51 per cent decrease in crime (showing similar results to the previous study) (Welsh and Farrington 2009).

Public Attitudes

There is a general lack of research into public attitudes towards CCTV. Public opinion tends to be incorporated into surveys as an add-on or afterthought. On the rare occasion when public opinion is the main focus of studies, it is often the methodology that is discussed at length, rather than the issues that arise with regard to the public. There is also a discrepancy between the results of amateur and professional surveys, focusing on public support for CCTV. For example, Ditton (1999, pp.201–24) found that amateur surveys usually find 90 per cent in favour of CCTV, and those carried out by professional researchers find around 60–70 per cent acceptability. He argued that sampling is a major factor in this regard, as well as the design of the survey or questionnaire. He provided a couple of examples of similar surveys with different results, conducted in the same area at around the same time, to highlight his explanation. Ditton also demonstrated that if questions are placed in a wider context of pro-CCTV questions the results differ considerably from those placed in the context of a survey on civil liberties (rather than a survey on crime control).

A prime example of this is a recent (2010) survey from the CCTV User Group (commissioned by the CCTV User Group and undertaken by TNS Research International) – 'An Independent Public Opinion Survey on the Use and Value of CCTV in Public Areas'. They surveyed over 1000 respondents (in their own homes, using multimedia technologies). Some of the results were as follows:

- 90% of survey respondents support the use of Public Area CCTV by Local Authorities and Public Bodies.
- 82% believe CCTV saves money by reducing Police and Court time.
- 80% of respondents believe that clearly visible CCTV managed by Local Authorities and Public Bodies does not infringe on their privacy rights.
- 76% consider there is the right amount or too few Public Area CCTV cameras currently operating.
- 71% believe that CCTV in public areas makes them feel safer and reduces crime.
- 70% are against any removal of CCTV cameras by public bodies to meet Government budget cuts.
- 63% believe that Crime and Disorder would increase if CCTV was removed in their area.
- 61% are against any reduction in the monitoring hours of CCTV by Local Authorities even considering the current economic climate.
- 70% are in favour of the continuing use of vehicle Number Plate recognition technology.

Arguably, the wording of the questions showed considerable bias, as well as being potentially leading. For example, one the questions went as follows: 'The use of CCTV in public areas assists in the 'post incident' identification of offenders or witnesses, and provides useful evidence of their movements' to which respondents had to state whether they agreed, disagreed, or didn't know. This sort of question leaves little room to disagree (terms such as 'useful evidence' lead the respondent down a certain path). Another question asked: 'The images captured by CCTV result in more frequent guilty pleas and savings in Police and court time as they provide incontrovertible evidence of any incident' to which respondents again had to state whether they agreed, disagreed, or didn't know. This question is once again leading – phrases such as 'incontrovertible evidence' leave little room for disagreement.

A study by Fyfe and Bannister (1996) mentioned the issue of public acceptance of CCTV (although this is included in a research study focused on public spaces and access entitlement). They looked at levels of public resistance to public space

CCTV systems and argued that support for these systems seems mainly to arise from a belief that they reduce crime and enhance public safety. Williams and Johnstone (2000) conducted a further small-scale survey, focusing on attitudes towards CCTV and its perceived effectiveness in tackling crime. They asked respondents to think about these issues in relation to crime prevention more generally, rather than as an isolated method. A greater number of police officers and increasing investment in activities for teenagers proved to be as popular as installing and using CCTV systems in the town centre, when respondents were asked to think about ways of cutting crime. More recently, Gill et al. (2007) conducted a study investigating public perceptions of CCTV in residential setting. This area of research has been particularly limited (even in comparison to the lack of research into public attitudes generally) – with the majority conducted in town and city centres, rather than residential areas. This study focused mainly on fear of crime and feelings of safety amongst residents following the installation of CCTV systems. The findings showed that support for CCTV decreased following the installation of CCTV systems due to little or no change in residents' fear of crime and feelings of safety.

Some research suggests that once the public enter into a debate on the efficacy and usefulness of CCTV (and they have the chance to voice their opinions and hear those of others), there is less unconditional support for the systems (see for example, Davies 1996). When the public are not engaged with CCTV they tend to welcome the technology and perceive it to be effective. It has been found that 'there is a strong desire to make it [CCTV] work' by the public (Hood 2003, p.251). In 2005, the Home Office funded a series of public attitude surveys (forming part of their National CCTV Evaluation (Spriggs 2005). These surveys were conducted in one town, two city centres and nine residential areas, prior to the installation of CCTV systems. The research looked at issues of: levels of victimisation, fear of crime, feelings of safety, and levels of avoidance of particular areas, awareness of existing CCTV systems, beliefs about capabilities of CCTV, support for CCTV, and privacy concerns. This research showed some interesting results regarding public knowledge of CCTV systems and their capabilities with 30 per cent believing that images from CCTV cameras are watched all the time. Research from Honess and Charman (mentioned previously in this chapter) also found a lack of awareness of the capabilities of CCTV systems and their functions. They recommended a greater level of public consultation to tackle this issue. I agree and argue strongly that any consultation open to the public on CCTV would be an improvement on the current situation. This is not just an argument from someone who believes that the public should be involved in the policy consultation process on science and technology, although I would also attest to that. It is an argument for thinking through the issues. Public consultation and engagement can mean that the uses and usefulness of a certain technology are thoroughly debated and thought through. This would certainly seem to be beneficial with regard to CCTV, when the empirical evidence often stands in contrast to the claims made about the technology by the authorities installing them.

Public Consultations and Discourse Surrounding CCTV

Consultations at the National Level

Recent consultations from the Home Office surrounding surveillance technologies and communications/information technologies have been:

- Interception of Communications in the UK consultation (1999).
- Public consultation on Draft Code of Practice on Communications Data (2001).
- Entitlement Cards and Identity Fraud consultation (2002).
- Access to Communications Data- respecting privacy and protecting the public from crime consultation (2003).
- Legislation on Identity Cards consultation (2004).
- Revised Statutory Code for Acquisition and Disclosure of Communications Data consultation (2006).
- National Identity Scheme: Delivery Plan 2008 (2008).
- Compulsory Identity Cards for Foreign Nationals – Consultation on Code of Practice (2008).
- Revised Interception of communications code of practice (2010).
- Amendments to the Anti-Terrorism Crime and Security Act 2001 (2010).
- Regulation of investigatory powers act 2000: proposed amendments effecting lawful interception (2010).
- A code of practice relating to surveillance cameras (2011).

A code of practice relating to surveillance cameras (part of the Protection of Freedoms Bill) was open for just under three months. This consultation received only 16 responses from members of the public, from a total of 107 responses. The consultation was described as seeking: 'views on the content of a new code of practice on the use of closed circuit television (CCTV) systems and other similar surveillance camera systems. The aim is to develop a code of practice for those using these systems, to help them to decide if using them is necessary, and how to get the best out of the systems and also for members of the public, to inform them about how these systems should be used and how to find out more about them. … We welcome comments from owners and operators of CCTV and automatic number plate recognition systems and cameras, including government, local authorities, police forces and businesses, and suppliers of such equipment and the public' (Home Office 2011). At first glance, this consultation is seeking to address an important issue – are the cameras necessary? However, reading between the lines may suggest a worrying emphasis; one of a need to 'inform the public' and to raise public acceptance levels of already installed systems (i.e. to inform of how they are being used, rather than to question their existence in the first place).

The Home Secretary (2011) stated that:

> We do not intend therefore, that anything in our proposals should hamper the ability of the law enforcement agencies or any other organisation, to use such technology as necessary to prevent or detect crime, or otherwise help to ensure the safety and security of individuals. What is important is that such use is reasonable, justifiable and transparent so that citizens in turn, feel properly informed about, and able to support, the security measures that are in place.

This resounding statement is reminiscent of earlier Home Office discourse surrounding CCTV. The question of whether CCTV actually detects or prevents crime is not asked, nor whether it actually ensures safety and security. These are provided as a given. Instead the code of practice suggests that transparency is important. Rather than wanting to engage the public in discussions over whether to use CCTV, how it should be used, what other methods might be available and so on, the code of practice equates transparency with acceptance (the public simply need to be informed and then they will accept the technology and the uses to which it is put).

The Protection of Freedoms Bill also states:

> Our approach to establishing a new regulatory framework is therefore intended to provide a means through which public confidence in CCTV, ANPR, and other such systems, is improved.

Again, the argument rests on the idea that decisions can be made in the area of regulation, which will in turn raise public confidence in the listed surveillance systems. The aim is one of raising public acceptance of the technology – not to question its fundamental basis or justification. In the final chapter of this book, I come back to this problem of a lack of public engagement in more detail.

Discourse

The House of Lords Select Committee report on 'Digital Images' (1998) argued for the importance of igniting and maintaining public support for CCTV systems, stating that 'public acceptance is based on a limited, and partly inaccurate knowledge of the functions and capabilities of CCTV systems' (section 4.8). They argued that a lack of public understanding and debate could potentially undermine the development and diffusion of CCTV and stated: 'We want to see public acceptance of surveillance … this is more likely to be the case if there is a wider public debate on the issues involved, and we consider that the government should provide such a debate' (section 4.22). This rhetoric is worryingly reminiscent of a Public Understanding of Science approach – that if only the public understood a little more about science (and in this case, technology) they would learn to love and appreciate it, and eventually support it (covered in more detail in Chapter 2).

The discourse starts from a point of seeking a high level of acceptance of CCTV, rather than really opening up the issue of surveillance to public debate.

The dominant political discourse surrounding the use of video surveillance in the UK has been that 'CCTV works' (see for example, Home Office 1996). CCTV has been presented in a positive light in the 'fight against crime', with the Metropolitan Police utilising the phrase 'CCTV systems are an important weapon in the fight against crime'. This phrase is also found repeatedly in newspaper articles including quotes from politicians (I go into this in more detail in the media analysis at the end of this chapter). Immediately prior to the general election in 2010, David Cameron, despite having already been quoted saying he wanted to sweep away the 'whole rotten edifice [of] Labour's surveillance state' also said that the Conservative Party supports CCTV in its role as a 'valuable tool in the fight against crime' and that the Party did not plan on reducing the number of CCTV cameras in the UK. However, he added that 80 per cent of cameras do not have a high enough footage quality to be used for prosecution purposes (BBC News 28 April 2010).

The 2007 National CCTV Strategy was described as follows:

> A series of workshops were held to understand the current CCTV infrastructure, its use and where the main issues and problems lie from the perspective of key stakeholders and interested parties. The workshops were an opportunity for users of CCTV and stakeholders to air their experiences and views on current public space CCTV. The workshops were organised to bring similarly interested groups together as part of the consultation. These included representatives from the following stakeholder groups: Serious crime, transport, government departments, the criminal justice system, technology and town centre CCTV groups. (p.9)

Civil liberties representatives/groups and the public were not asked to participate in the workshops or consultation. The consultation did not include an opportunity for debate on the relevance of and supposed need for CCTV in the UK, instead centring on how to improve the technology. In an interview I conducted in 2010 for the purpose of a project, I asked one of the people leading the consultation why the public were not involved (name and institution anonymised) and they replied that the strategy was about 'putting right what we'd already got wrong' (i.e. making the CCTV cameras already installed become useful and effective). When asked why the public were not invited to participate in the consultation, the response was that 'the public were represented by local government'. In any case, the strategy document stated that: 'CCTV in the UK enjoys significant public support and year on year fear of crime surveys states that the public feels safer due to the presence of CCTV'. The public had therefore already been constructed as a group who support the use of CCTV.

Following the recent announcement of the Protection of Freedoms Act 2012, the Minister of State for Crime Prevention, Jeremy Browne, stated that with this Act the Government 'intend[s] to ensure that surveillance camera systems continue to be an important tool available to communities to help tackle crime and prevent terrorism whilst balancing public safety objectives with the individual's right to privacy' (Browne 30 September 2012).

The balancing of security/crime prevention/combating terrorism and privacy/civil liberties is found often in political and media discourse surrounding video surveillance technologies. It is often cited to be a zero sum game; that one must be traded in order to gain the other (and that this works). Daniel Solove makes a detailed argument against this line of thought in his book *Nothing to Hide: The false tradeoff between security and privacy*. Solove essentially argues that we are often forced to choose between one and the other, and that this does not make sense. Losing personal privacy does not equate directly to a gain in terms of security or safety. Examples of this discourse are rife in UK political and media rhetoric. To take one example, in the immediate aftermath of the London Underground bombings in 2005, there were repeated calls for increased security on the London Underground (despite the fact that the cameras that were already installed could not and did not stop the incident from happening). The common argument found was that the public would need to accept a higher level of security and surveillance equipment (with a focus on CCTV cameras) in order to gain safety. To illustrate this point, the following come from a variety of newspapers at the time:

> The Government should immediately look into ways of toughening up security on London's transport system ... At the very least this should mean CCTV cameras on tube trains. (*The Independent* 23 July 2005)

> After today, I expect the travelling public will be more prepared to put up with a greater level of surveillance. (*The Times* 8 July 2005)

> More than three in four say the best way to prevent further attacks on London and other cities is to increase the number of CCTV cameras. (*Mail on Sunday* 10 July 2005)

> I've no doubt that in the short term it's going to get somewhat uncomfortable, perhaps even painful, especially where civil liberties are concerned and especially for British Muslims. But the fact is that this short-term pain or discomfort is a price worth paying for long-term security and community harmony. (Shahid Malik, Labour MP to Dewsbury quoted in *The Sunday Times* 17 July 2005)

However, Tony Blair is quoted in one article as saying 'all the surveillance in the world could not have stopped the London bombings' (*The Mirror* 11 July 2005).[1]

1 For a more detailed analysis of the representation of CCTV in the print media after the London Underground bombings, see: Kroener, I. (2013) 'Caught on Camera' *Surveillance and Society* 11(1/2)

This point is equally true of the 2004 train bombings in Madrid, Spain. I will return to this point in Chapter 6, which focuses on international comparisons.

The Local Promotion of CCTV

Discourse at the local level

Previous studies of local authority discourse surrounding CCTV have found that it is linked to revitalising town centres and to encourage shopping (Graham et al. 1996). In my own research, I have also found CCTV to be linked to consumption and making the city seem like a nicer place to be (at least for those who are going to spend money there). Citizenship and inclusion are important themes in discourse surrounding CCTV; and in this way and many others, CCTV is a very 'them' and 'us' technology: there for law-abiding citizens and consumers, and against those who do not conform to these categories. For example, Cambridge City Council state that:

> The purpose of our CCTV system is to make Cambridge a safer and more welcoming place at any time of the day or night – allowing all citizens and visitors the opportunity to participate fully and without fear in the life of the city.

Peterborough City Council cite similar reasons for the installation and use of a CCTV system in the city centre stating the purpose as:

> To help provide a safer and more secure, user-friendly public environment in the City Centre and other areas covered by the scheme for the benefit of the whole community, including those who live, work, trade, visit and enjoy the facilities of these areas.

Discourse tends to emphasise participation in civil society.

The phrase 'people now feel safer when they're out and about' is found repeatedly on local authority websites. Despite numerous studies questioning the effectiveness of CCTV (detailed previously), the technology finds its use in making people 'feel' safer. Discourse focuses on not only tackling crime but also fear of crime. For example, Southwark Council discuss the reasons for the use of video surveillance in the city centre as not only for 'crime prevention and detection' but also to 'raise public confidence and feelings of safety'.

The idea of constantly being watched or surveilled is found often in local authority discourse. The discourse focuses on the positive impact that constant surveillance has on the law-abiding citizen (CCTV is therefore a protective technology for those who are the desirable citizens or consumers in the city centre). For example, Birmingham Council use the following extract on their website:

> In safe hands. Ever felt like you're being watched? It's not surprising! Wherever you go these days, in shopping complexes, railway stations, car parks even just

> sitting by a fountain – you are often being watched. Closely … The Centre has fibre cable links to both the West Midlands and British Transport Police. So, the next time you're reading your paper by a fountain, be reassured about who's watching over you!

Other councils focus on privacy concerns or issues that may arise from the installation and use of CCTV systems. For example, Derby City Council state that:

> Arguably, Closed Circuit Television (CCTV) is one of the most powerful tools to be developed during recent years to assist with efforts to combat crime and disorder whilst enhancing community safety. Equally, it may be regarded by some as the most potent infringement of people's liberty. If users, owners and managers of such systems are to command the respect and support of the general public, the systems must not only be used with the utmost probity at all times, they must be used in a manner which stands up to scrutiny and is accountable to the very people they are aiming to protect. The Council is committed to the belief that everyone has the right to respect for his or her private and family life.

Not only are privacy issues mentioned, but data protection and the use of CCTV images is often detailed on local authority websites. They clearly set out the uses for which images captured on CCTV cameras can be used, details about processing and storage, and on occasion retention periods. Many of the local authority websites also detail whom to contact with questions regarding CCTV and data protection. However, as is the case in general with CCTV, it is not necessarily the lack of legislation that causes problems, but rather that this legislation is only there on paper (as such). It is often not followed up by regulation or enforcement. A few years ago I undertook empirical work for a research project focusing on the use of signage by local authorities (and whether this was in keeping with the Data Protection Act) and found it to be distinctly lacking. On occasion a sign would be put up underneath a CCTV camera, simply saying 'CCTV in Operation'. These sorts of signs are not in line with data protection legislation, which states that CCTV signs must be clearly visible and readable, and also provide details of who is operating the system and their contact details.

Consultations at the local level

With regard to CCTV there are examples of public consultation at the local authority level. For example, Barnet Council currently (October 2013) has an open consultation seeking views on a new layout for CCTV cameras across the Borough. Other councils provide publicly available information on their websites. Wycombe District Council, for example, use the online space to provide access to minutes of CCTV meetings, recent CCTV newsletters, annual public meetings, the operational parameters of the system, details of the lay public committee (which provides oversight of the system) and the means through which members of the public can request access to CCTV footage. However, there are also a number of examples where the only information available with regard to CCTV is a list of the benefits of the use of video surveillance for crime prevention purposes, with no opportunity for the public to express their views on the use of these systems at the local level.

On closer inspection, consultations at the local authority level still maintain an overriding discourse that the effectiveness of using CCTV to prevent crime is without question. The current consultation by Barnet Council includes an online questionnaire. The consultation document itself begins with the phrase:

> CCTV Cameras are a valuable tool to the London Borough of Barnet to help keep our communities safe, reduce crime rates and provide evidence for the London Metropolitan Police to secure criminal convictions in the courts.

The document continues with a list of the reasons why CCTV has been installed, followed by a list of the location of cameras. There is no evidence provided that suggests that CCTV may not be effective in terms of its aims. The online questionnaire includes questions such as:

> In your opinion, which of the following activities is CCTV most effective in reducing?

This question is followed by a list that includes: anti-social behaviour, theft from motor vehicle, violence against the person and so on. There is also an option to answer 'don't think that CCTV is effective in deterring any of the above'. The questionnaire does include one question asking whether the respondent would like to see the removal of any cameras that are currently installed, however it is not clear whether the removal would simply entail moving the camera from one location to another.

As mentioned above, there are examples of lay oversight committees, where members of the public are invited to yearly meetings to discuss the use of CCTV in the local area. However, to use the example of Wycombe District Council, the minutes of the last committee meeting suggest a lack of effective dialogue between the experts invited to take part in the meeting and the lay members of the advisory panel. Furthermore, one the aims of the lay advisory panel is stated as being:

> To assume the role of critical friend of, and give the best possible advice to, the Council on (a) the operation of the CCTV System, including but not restricted to how it might be improved in order to achieve its primary purposes of protecting the rights of the public by detecting and deterring crime and unlawful behaviour and (b) how best to inform the public of its operation and efficacy.

Rather than opening up a meaningful dialogue about the uses of CCTV (including whether it should be used at all, or whether there are other potentially more effective means of crime prevention/detection), the purpose of this lay oversight committee is established from the outset as one of being a 'friend' to CCTV (whether critical, or not). There is also no opening to discuss the possibility that CCTV may not achieve its primary purpose in practice. The discourse is one of crime prevention and 'efficacy' – and the aim of opening up the operation of CCTV systems to members of the lay panel becomes one of 'informing' the public about the benefits of using CCTV.

Engagement

Public consultation on CCTV in Britain is distinctly lacking. This is also the case in terms of other forms of public engagement (citizen juries, focus groups, and so on). In recent years, and in other areas of science and technology, particularly those relevant to the environment, there has been a significant increase in public engagement activities. Methods used to engage the public include: deliberative polling, citizens' juries, focus groups, consensus conferences, stakeholder dialogues, Internet dialogues, and deliberative mapping. It has been found that public engagement has been held back in corporate and private settings, which we could also take to include security and surveillance domains, and has been most prevalent with regard to environmental issues (Wilsdon and Willis 2004).

This may be unsurprising. As I have already demonstrated in this chapter, the discourse around CCTV is one that promotes CCTV as a security technology, designed and implemented to provide safety to a passive public in need of protection. If its uses are already so clearly defined, and the role of the public likewise, then what is there to discuss? In the 2007 National CCTV Strategy consultation, the public were not asked to participate (it was assumed that they were represented by local government).

I feel strongly that the public should be engaged on these issues. Not least because it is public money that is being spent on these systems (and a lot of it). The discussion around the use of CCTV needs to include those who are affected by how it is used, to what extent it is being used, and how it potentially takes funding away from other public services:

> The purpose is to hold science and technology answerable, with the utmost seriousness, to the fundamental questions of democratic politics – questions that have fallen into disuse through modernity's long commitment to treating science as a realm apart in its ability to cater for society's needs: Who is making the choices that govern lives? On whose behalf? According to whose definitions of the good? With what rights of representation? And in which forums? (Jasanoff 2007)

In the next section, I provide the results of an empirical study I conducted in 2009. In this study I focused on the planned installation of CCTV onto two local authority run estates in London. In this small-scale study I wanted to assess some of the questions asked by Jasanoff above: Who is making the decisions about where to install CCTV systems? Have the public/residents been consulted? If so, in which manner and through what means? If not, who is defining where and how it should be used?

The Digital Bridge Project

As has been shown throughout this book, much of the empirical evidence points towards CCTV systems that are ineffective, or at least not very effective, in preventing crime. Advocates of these systems argue that they can be useful in prosecuting people after the event. CCTV is certainly used to aid investigations (although its usefulness in this context is also often overstated). However, that is not really the point in question. The discourse surrounding CCTV is that it 'works' (and presumably this should refer to the purposes for which it is installed: to increase public safety and to cut crime). The public have had little say in how, where and when these systems are installed. I believe that there is often another method or solution to increase public safety, or at least that other methods should be discussed, taking into account local situations and needs. The following section looks at a specific case study in London.

On 8 May 2006 the Digital Bridge project was launched in Shoreditch, London by the Minister for London and Neighbourhood Renewal, Jim Fitzpatrick. This project planned to bring broadband to two estates in Hackney, East London: Charles Square and Haberdasher Estates, through a PC on TV service. Also included in the project proposal was the installation of CCTV cameras (known as 'community webcams'). The proposal was that residents could watch images from these CCTV cameras on the new Community Safety Channel (through the PC on TV service). The project was awarded the title 'Innovator of the Year for 2005' by London Connect.

The service was delivered by the regeneration agency – the Shoreditch Trust – in partnership with Video Networks Ltd. and received funding from the Office of the Deputy Prime Minister (now the Department for Communities and Local Government), the European Union and private investment, totalling £12 million.

Despite winning awards prior to its inception and glowing political endorsements, such as:

> Digital Bridge is an immensely important project … . It will allow people to report graffiti online … And to be able to do that very, very easily through the technology. It's going to be fantastic for people and I think it is a really exciting project. And if it works as well as we hope then we can see that project extending in many different parts of the country. (Tony Blair, Prime Minister)

> We are interested in how local authorities across the country can use the internet and web portals to allow people to … receive the services they use. I have looked at Shoreditch … at the Digital Bridge that allows people to alert residents as

> events happen, and residents to alert them about abandoned cars, about graffiti, about vandalism. (Gordon Brown, Chancellor of the Exchequer)

This project never came to fruition. The cameras were installed but are not used. The broadband expansion failed. I argue that the project ultimately failed (and somewhat importantly was not explicitly stated to have failed in any press releases, media coverage, or government documentation), at least in part, due to a lack of public engagement. Although the pilot results were published and showed promise in terms of potential uptake of the service, the results of empirical research show that this service was not taken up by residents and is known to have failed. My research shows that the service was ultimately too expensive for the lower income households on the estates. Furthermore, residents were not engaged with the project prior to the installation of the cameras, or the attempted installation of broadband onto the estates.

The pilot results for the Digital Bridge project in Shoreditch stated that:

- 70% regularly accessed and used the live webcam network.
- 46% of users were reporting crime through the Action Maps (an interactive application detailing where crime has been reported and where it is being cleared).

However, the empirical work I conducted on the estates showed that the service was not taken up by residents for two reasons: broadband does not work on these estates (it had been trialled by a major telecommunications company three years earlier and failed), and the cost of the service was too high for these lower income households (£3.50/month). Perhaps even more importantly, the majority of residents did not know about the project beforehand, or even once the cameras had been installed. Had the Shoreditch Trust conducted the research which they claimed to have done, they would have found out, prior to spending £12 million that the project was destined to fail.

This is a prime example of the way in which CCTV has developed in the UK: it is a political tool used by politicians and local authorities to be seen to be doing something about crime, it is used as a silver bullet, a 'technological fix', a technology utilised and widely disseminated without public consultation, and a technology situated outside the socio-economic context. Although my empirical research amounted to a small-scale study, and was never designed to be representative (in terms of general views about CCTV, rates of public acceptance, or attitudes towards CCTV) I believe it highlights some important issues around CCTV and how it is used.

Of the 35 respondents to my questionnaire, only one was invited to participate in a focus group prior to the installation of cameras on the estate. Of the 34 residents who had not been asked to participate in a focus group, 30 stated that they would have liked to be consulted prior to the installation of the cameras. When asked why they did not subscribe to the Digital Bridge service, the residents answered that they: 'have not heard of it', 'don't know much about it', 'no one offered it to me' or 'too expensive/cannot afford it'. One respondent also commented that it is not in operation anymore, claiming that the Digital Bridge was removed over a year ago. They said:

> Digital Bridge failed because it did not do its market research properly. Because of this area's proximity to the city and/or its high technology, we have NEVER been able to receive a decent TV picture without cable. Even in the 1970s everyone had a form of cable. Digital Bridge therefore was an additional expense on top of our normal cable expenditure. ALSO it would only operate on a BT line. Here again, most people had a cable package that included phone, broadband and TV. BT were their usual unhelpful selves and insisted on a new line number if we changed. Well, why bother! Another large sum of community money lined someone's pockets.

In response to questions about the Community Safety channel, residents answered that: 'in theory a crime could be detected and reported', 'only if actively promoted and recordings are used', 'if it makes people feel more secure it can only be good', 'safety crime reduction, better type of policing', 'only if people do something with it instead of just using it as another TV channel to view', 'a method of crime prevention'.

Concerns raised over the Community Safety channel included: 'intrusion of privacy', 'no one knows about it', 'invades privacy', 'don't know if images are strong enough to identify anyone engaging in criminal activity', 'if so many cameras were put in that they could be used to look inside people's flats then that would be an invasion of privacy', 'gross invasion of privacy – particularly so in the context of overall government collecting of personal data'. In the context of concerns over the channel, the residents' views were mixed. For some the issue of whether the channel and cameras were effective and a justifiable (or even attainable) expense was a matter for concern, and for others it was the privacy implications that were the major issue.

I included a space for other comments, which resulted in a range of views. One person stated: 'I did subscribe to digital bridge when it was first installed and kept it until just before the cameras were taken down. No one told us they were to be taken down we just found out somehow'. This was the same person who had described the lack of market research causing the failure of the project. It was only through this respondent and one other that I found out that the project had failed. There was no information available to this effect from the Shoreditch Trust. The other respondent who had knowledge of the project and its failure stated that: 'I only heard about Digital Bridge through the media. I also heard that it had now been discontinued. No information was given to residents about the scheme or an update on what has happened to it'. I could not find any press articles that detailed the results, or the failure, of the project.

One of the comments that really stood out to me from this research study was from a resident who stated: 'CCTV breeds a culture of fear. Civil courage, rather than CCTV/surveillance should be encouraged'. As I discussed earlier in the introduction to this book, there is an issue regarding whether CCTV should be used at all. Instead of providing safety and security, does it instead promote feelings of insecurity, and a loss of feelings of responsibility towards others?

For the purposes of this research study I conducted interviews with the residents. It was clear that CCTV provokes varied feelings in people. For some the cameras made them feel safer. For others they made them feel 'worried'. One respondent argued that:

> Only that it's presented as being a kind of magic bullet and it isn't, so that's a concern … because every time someone thinks you can just refer back to something and that'll give you the answer and therefore solves the problem … then, there could always be the problem … the problem's already been done, as such.

In terms of public discussion or consultation on CCTV in general, one interviewee suggested there has been an exclusion of certain members of society, or certain groups:

> [We're] excluded on purpose. They only include people who are going to support … especially with public debate, they never seem to go to anybody who'll oppose what they're saying.

I asked at this point whether any members of the public were involved in the debate, and the interviewee replied:

> People are, just not the people who are actually involved in it, or who it might affect in any way.

As I mentioned earlier, this was a small-scale study, not seeking any representative results in terms of British attitudes towards CCTV or public consultation, but I strongly believe that even a small study can point towards important issues that need addressing. The interviewee quoted above states that the 'right' people are not involved. That the people who are actually affected by the technology are not engaged in the debate or policy consultation process. At a local level this can have a hugely negative impact. The amount of money spent (and wasted) on the Digital Bridge project is, for me at least, extremely sad. Money was poured into a project for political reasons, rather than seeking to address some of the actual issues that are found in socio-economically deprived local authority estates in London. To conclude this section, I will share one final comment from an 87-year-old resident. When asked how security and safety on the estate could be improved, the interviewee responded:

> I come home sometimes and there's boys from the local school sitting on my mat playing cards and I have to say 'excuse me' to get into my flat, and I wonder whether one day they're going to shove me in and grab my handbag but it hasn't happened yet. It will be nice when there's locks on the doors so as no one can get up to my door unless they're visiting somebody – there'll be a door either end which we'll have keys to, as long as everybody shuts that you'll only get the genuine postman and if someone you're expecting you'll let them in. It would be much nicer if it were locked off. Although I don't know how it's going to keep everyone out as people are a bit absent minded about closing the door …

> I mean, why didn't they get the keys ready before they put the gates up? They've been there for two years but there are no keys.

> I mean we're not tackling it [crime] in the right way are we? I mean, why don't they ask the people?

I believe this project represents a failing of policy-makers and politicians to engage the public with CCTV. For a variety of reasons, it is a great shame that they didn't. Possibly the biggest shame is the huge amount of public money that was spent on a system that was never going to work. Had the residents been consulted during the planning stages of this project, the problems with installing broadband into the area would have come to light. Equally important though is the lack of dialogue about local needs in terms of crime prevention and security. Public engagement at a local level would highlight the smaller changes that could be made that would increase residents' safety (for example, putting locks in entrance gates as one respondent suggested), rather than a swoop towards the grander techno-fix of installing a CCTV system onto the two estates.

CCTV and the Media

So far, this chapter has focused on the discourse surrounding CCTV, empirical studies into the effectiveness of CCTV, public acceptance and public attitudes, as well as a discussion of the lack of public engagement with regard to CCTV systems. I presented the results of a small-scale empirical study highlighting the problems with a lack of public engagement in this area. This next section analyses the representation of CCTV in the British print media from 1992 to 2008. Although I do not subscribe to a linear model in relation to media discourse directly infiltrating the readers' perceptions of issues, I do subscribe to a two way process in terms of the media and the audience. I agree with Chapman (2005, p.1) when she argues that 'the media make the world even as the world makes the media'.

Media Discourse

It has been argued by some that we are living in an age of 'moral panic', which can be defined as 'a high level of concern over the behaviour of a certain group or category of people and … an increased level of hostility toward the group or category perceived as a threat' (Thompson 1998, p.9). In recent times, there have been a number of groups that fit this category: terrorists, suicide bombers, and those who abduct children; where CCTV has been promoted as a solution (or at least part of the solution) to incidents reoccurring and the ability to stop these groups of people (known as 'deviants' or 'others' in the traditional sociological literature).

This age of moral panic is coupled with the emergence of what some suggest is a 'risk society'. Risk is not a new concept. Luhmann (1993, p.9) describes the term as first emerging with reference to maritime voyage (and to describe the perils that this offered) during the transitional period between the late middle ages and the early modern era. Risk at this time was conceptualised as resulting from natural events. Giddens argues that this changed with the onset of Modernity. During this time, an idea evolved that 'the key to human progress and social order is objective knowledge of the world through scientific exploration and rational thinking. It assumes that the social and natural worlds follow laws that may be measured, calculated and therefore predicted' (Giddens 1998, p.94). This idea of being able to measure the social and natural world began to be coupled with the notion that risk itself could also be calculated and measured. The notion of risk was also changing during this time to include human-made events.

Beck (2006) argues that we are now living in a 'risk society', a modern society that is 'increasingly occupied with debating, preventing and managing risks that it itself has produced'. Luhmann (1993, p.83) suggests that 'more than any other single factor, the immense expansion of technological possibilities has contributed to drawing public attention to the risks involved'. Risk is therefore something that is linked to technological development, and the use of technology. However, technology is also seen as something that can decrease risk (for example, surveillance technologies used to combat terrorism, or fear of terrorism). Douglas and Wildavsky (1982, p.32) argue that technological development has brought with it an expectation that risks and problems can be solved via technology, stating 'low technology originally set the lower normal expectancies, but high technology has given us now the hope that any present level of bad things may be lowered'. These sorts of expectations (coupled with the idea of technology as progress) have given rise to a modern ideology of a 'techno-fix'; that social problems can be solved via the use of technology.

When it comes to media coverage of crime, Reiner (2002, p.380) states that 'crime narratives and representations are, and have always been, a prominent part of the content of all mass media'. He argues that deviance becomes 'the defining characteristic of what journalists regard as newsworthy' (2002, p.305). There are also various studies that pick up on the idea of moral panics and the notion of 'folk devils', referring to a group of persons or a person becoming defined as 'a threat to societal values or interests; its nature is presented in a stylized and stereotypical fashion by the mass media' (Cohen 1973, p.9), to analyse media content in terms of 'evil doers' and 'innocents' (Lee 1984), or 'them' and 'us' (Finn and McCahill 2010). This idea of 'innocent others' is also discussed in a 2003 study by Chibnall, where he finds that stories containing incidents against this group as being most newsworthy. Altheide (2002) uses the example of the news coverage on crime – and in particular the victimisation of children – to develop an argument that a discourse of fear has become particularly prevalent in modern societies. Fear, he argues, has become a central feature of the framing process, and that this discourse of fear that arises has infiltrated media coverage, as well as becoming a general frame of public discourse through which people understand and construct social situations and the world around them.

Of course, there is not a linear transmission of information from the media directly into a formulation of thought on the part of the audience. This dominant model of science communication (Hilgartner 1990) is outdated and incorrect. However, media content does have some influence on the public (and this is a two-way process, with public opinion also influencing media coverage). Gurevitch and Levy (1985, p.19) describe the media as 'a site on which various social groups, institutions and ideologies struggle over the definition and construction of social reality'. Gamson (1988) adds to this idea with his argument that news discourse feeds into the framing of public policy issues and helps to shape public debate and opinion on issues. Pan and Kosicki (1993, p.64) discuss the importance of news discourse, arguing its influence and direct relevance to public policy making, stating that:

> Within the realm of news discourse, causal reasoning is often present, including causal attributions of the roots of a problem, inferences about the responsibility for treatment of the problem as well as appealing to higher level principles in framing an issue and in weighing various policy options.

In terms of framing, this occurs in the media when journalists 'select some aspects of a perceived reality and make them more salient in a communicating text, in such a way as to promote a particular problem definition, moral evaluation, and/or treatment recommendation for the item described' (Entman 1993, p.52). Journalists do this by the 'presence or absence of certain key words, stock phrases, stereotyped images, sources of information, and sentences that provide thematically reinforcing clusters of facts or judgements'. Frames can therefore 'provide context that is communicated with the text, and also can shape the way text is received' (ibid.).

This does not necessarily mean that information is directly received by an audience, rather that an audience is capable of differing interpretations of information presented (Neuman et al. 1992). Frames can therefore be defined as 'interpretive packages', which provide meaning and context to an issue. They become a 'central organising idea or frame, for making sense of relevant events, suggesting what is at issue' (Gamson and Modigliani 1989).

Media Representations of CCTV

There is currently a growing literature on surveillance and the media (see for example, Coleman and Sim 2000, Doyle 2003, Kammerer 2004, Albrechtslund 2008, and Finn and McCahill 2010). There is also a smaller literature concentrating on CCTV and the media. Research conducted by Norris and Armstrong (1999), McCahill (2002), and McCahill and Norris (2002) analysed representations of CCTV after the Jamie Bulger case, shows that media reportage focusing on the introduction of CCTV cameras at this time was almost entirely positive. Barnard-Wills (2011) analysed UK news media discourses of surveillance in general but also mentioned CCTV in relation to the London bombings. A more recent paper (Kroener 2013) focused on the representation of CCTV in newspapers in the immediate aftermath of the London Underground bombings. There is also a paper by Hier and Greenberg (2009), which analysed the framing of, and representation of CCTV by the media, but this is based in a Canadian context. There is certainly scope for further research in this area. The next section presents some preliminary findings from an exploratory scoping study of national newspapers in the period 1992–2008.

In my own research, I have found that mentions of CCTV in newspapers in Britain have steadily increased during the last few years.[2] Prior to this, there were a number of peak years in terms of coverage: 1993, 1994, 1995, 1998, 2000, 2002, and 2005. Since 2006 there has been a consistent increase in terms of articles mentioning or focusing on CCTV. In this section, I want to concentrate on the years prior to 2005 in the majority, when there were still fluctuations in the coverage of CCTV. I do also include coverage post-2005 but my main focus is before this date.

The last 20 years has seen a dramatic increase in the number of newspaper articles focusing on CCTV. The early 1990s saw *The Guardian* approach the increased use of the technology with criticism, whilst *The Times* depicted the developments in a largely positive light. The *Daily Mirror* and *Daily Express* also represented CCTV in a positive light during the 1990s. During the early 1990s, coverage tended to focus on the effectiveness of CCTV and the results of pilot schemes. In 1993, *The Guardian* reported some low quality footage hindering prosecutions. A year later both *The Guardian* and *The Times* began to promote the usefulness of CCTV in cutting crime using phrases such as 'success' and 'raising hopes'. David McLean, the junior Home Office Minister, was reported in *The Guardian* as saying that CCTV reduces shoplifting, car and drug related

2 I conducted a keyword search for 'CCTV' and 'Closed circuit television' across all national newspapers from the mid-1980s, using the database *LexisNexis*. As this was an exploratory qualitative study I did not code for certain words but wanted to leave the study open to explore the main themes arising in relation to CCTV. There is scope for a more in-depth content analysis of the key themes found during this period.

crimes, and Michael Howard, the Home Secretary announced £2 million CCTV competition for local authorities and estates. In late 1995, he was quoted in *The Times* (23 November) as saying:

> Criminals hate the cameras. They know that CCTV allows law-abiding people to reclaim their streets and public spaces.

This was reported at the same time that results of the first CCTV competition were announced. Terms such as 'spy cameras' and 'controversial spread' began to be used by *The Guardian* during 1995, signalling the start of criticism against CCTV by the newspaper (although this has not remained the case from this point onwards, which will be shown later in this analysis), at this point focusing mainly on issues of privacy, data protection, data security and effectiveness (*The Guardian* 22 March 1995; 21 September 1995). At the same time, *The Guardian* (30 January 1995) reported police backing of CCTV, and quoted David McLean as saying that CCTV is the 'friendly eye in the sky'. The paper also reported some criticism from academic evaluations over success rates and some 'uneasiness' in terms of public feelings towards 'spy capabilities' (ibid.). One report in *The Guardian* at the beginning of the year suggested that there was general public acceptance of the rapid growth of CCTV cameras, finding this to be at odds with the traditional British approach to crime:

> The lack of any marked reaction to the use of CCTV into public areas in 1994 came as a surprise to many inside and outside the security industry. 'Teams of journalists flocked out to see us', says Tim Johnson, security manager for Liverpool Town Centre, 'All of them expected to come back with Big Brother horror stories. But people in Liverpool simply welcome the scheme'. The reaction of people in Liverpool may be partly explained by the city's proximity to the scene of the James Bulger tragedy. 'After the Bulger outrage', says Fred Domenico, at Asprey in Bond Street, 'public feeling was Catch Them, and by association all criminals, by any means. Nevertheless in the UK we have traditionally taken a more laid-back approach'. (22 March 1995)

For the *Daily Mirror*, CCTV became a political issue during 1995, with the launch of an inquiry by the newspaper into the areas that received Home Office funding for camera installation. It reported that:

> Home Secretary Michael Howard was accused last night of pouring crime-busting cash into Tory heartlands – and snubbing trouble spots packed with Labour voters. (10 April 1995)

The article went on to argue that 'true-blue' areas (i.e. Conservative) received funding for CCTV when there was a low crime rate and cameras already installed, at the expense of more deprived inner city areas. The general tone towards CCTV was positive though, in terms of the effectiveness of the technology. This positive tone was echoed in the *Daily Express* during 1995, with a number of articles detailing the success of CCTV in reducing crime and reporting on funding for the installation of cameras throughout the country (3 August 1995; 29 November 1995).

There was a definite shift in coverage on CCTV in *The Guardian* during 1998. A critical look at the effects of CCTV on crime was taken, with one article detailing possible displacement effects and a rising fear of crime in places not covered by cameras, as well as referring to CCTV as 'Big Brother' (9 January 1998). Later in the year, *The Observer* raised concerns over civil liberties in relation to the proposed launch of CCTV with facial recognition capability in Newham, East London. It described a quote from a designer of the system as a 'chilling vision of the future' and reported:

> Gradually this sort of technology will cover the whole country. Then it will become unprofitable to commit crime anywhere. (11 October 1998)

The same development is described in *The Guardian* as 'spy camera[s]', once again raising concerns over possible civil liberties infringements (15 October 1998). *The Times* also offered some criticism of CCTV during the same year, reporting on the poor quality of footage for the purpose of identifying people (27 March 1998). The *Daily Mirror* did not cover the civil liberties angle during 1998, focusing instead on the success of CCTV in prosecuting criminals after being 'in view of CCTV cameras', 'caught on camera' (6 March 1998) and 'tracked by CCTV cameras' (26 August 1998).

A change in coverage in *The Times* occurred during 2000, with an increase in mentions of privacy and the implications of the technology. *The Guardian* also increasingly mentioned privacy and excessive surveillance. There were numerous mentions in *The Times* of snooping and surveillance, as well as references to Big Brother (10 January 2000; 11 January 2000; 20 January 2000). With reference to a new facial recognition technology – 'SatNet' – being used in conjunction with CCTV (used to match the faces of shoppers to known shoplifters on a database) *The Guardian* raised questions about privacy infringements and 'being watched' so closely. It also reported 'excessive surveillance' and a call from Liberty for a public debate on surveillance-related issues (10 February 2000). However, although the number of articles mentioning privacy implications or other negative effects of CCTV rose during the first half of the year, the majority of coverage continued to paint the use of video surveillance in a positive light.[3] During this time, neither the *Daily Mirror* nor the *Daily Express* focused on the negative aspects of CCTV. The

3 Between the 1 January 2000 and 1 July 2000, the *Times* contained thirty-four articles mentioning CCTV. Of these, five contained negative terminology such as 'snooping', 'surveillance' and 'Big Brother'. During this time, the *Guardian* contained fifty-nine articles mentioning CCTV. Of these, four articles depicted CCTV in a negative light, using terms such as 'excessive surveillance', 'monitored' and 'watched too closely'.

Daily Mirror started to report the checking of CCTV footage as a given, prior to knowing whether it will be of any use. For example, in relation to a murder in Ireland, it stated that police 'will be looking at every inch of CCTV footage to spot Julie and see who she was talking to' (5 February 2000). This reporting of CCTV being checked, prior to its usefulness being assessed, occurs numerous times – 'Police were checking CCTV footage from the club' (24 May 2000), 'Officers are studying CCTV footage for clues' (10 June 2000), '[Police] are searching film from Gloucester's CCTV for pictures of the rapist' (19 August 2000). The police going through CCTV footage then becomes newsworthy in itself.

In July 2000, the court case of the Brixton nail bomber[4] was covered by the *Guardian*, with the role of CCTV leading to his identification mentioned in three articles: 'Police tracked him down through a series of leads, including calls from members of the public who recognised him from CCTV footage released following the Brixton bombing' (2 July 2000). *The Times* mentioned CCTV and its role in this case in one article (1 July 2000). The *Daily Mirror* took a strong view on the role of CCTV in the case, reporting that 'Copeland was caught after a CCTV photograph of the Brixton bomber appeared in the press' (1 July 2000) and 'A few hours later, he was arrested at his home after being identified by a workmate who recognised a CCTV shot of him taken at Brixton' (ibid.). The *Daily Express* did not mention CCTV in relation to the case.

Also in July, *The Observer* reported possible privacy infringements from surveillance at work, through computers and the Internet, via CCTV, and loyalty cards (30 July 2000). A couple of months later, in September 2000, *The Times* focused on CCTV and the Human Rights Act, in relation to the launch of a 'Supervan' fitted with nine cameras on patrol in Westminster (19 September 2000). Privacy was also mentioned in an article in the *Daily Express* in September, reporting on the inappropriate use of technologies in the workplace; in the case of CCTV, installed to protect employees from attack. The article argued that 'we all have a right to privacy' (15 September 2000).

On 27 November 2000, 10 year old Damilola Taylor was stabbed outside his block of flats in Peckham, London. *The Times*, *The Guardian*, *Daily Express* and *Daily Mirror* all mentioned CCTV as recording his last journey (*The Guardian* 1 December 2000; 2 December 2000; 5 December 2000, *Daily Mirror* 1 December 2000; *Daily Express* 5 December 2000). *The Guardian* also reported that the children who were with Damilola Taylor shortly before he was stabbed had been identified through CCTV footage. Between 29 November and the end of the year (2000), *The Guardian* published 98 articles mentioning Damilola Taylor. Of these, CCTV was mentioned in nine. During this time, *The Times* published 71 articles mentioning Damilola Taylor, of which two mentioned CCTV. Within their report of CCTV footage showing Damilola's movements prior to his death, the *Daily Express* described the recording as

4 David Copeland, known as the 'Brixton nail bomber' carried out a 13-day bombing campaign in London during April 1999.

showing him in a 'carefree mood', using images with no sound to ascertain his feelings prior to the attack (5 December 2000). This placing of CCTV into the role of body language expert occurs again numerous times during coverage of the London Underground bombings, which I come back to later in this chapter.

In January 2002 *The Times* reported that street crime had risen dramatically (giving the specific example of Harrow) in London boroughs since 9/11, due to many police being involved in anti-terrorism duties, leading to a fall in on the beat policing (10 January). A couple of months later (9 March), *The Times* reported on 'Fortress Britain', detailing communities behind electric gates and gated developments. There was brief mention of CCTV, but *The Guardian* – in its reports of gated communities, during the first half of 2002 – was far more emphatic about the role of CCTV in enhancing security (30 January 2002). The *Daily Express* (4 April 2002) included an article in April focusing on data trails, satellites, purchasing habits, home and mobile phone records and CCTV, entitled: 'There's no hiding place from this army of spies'. This article was included at a time when there was little or no reference in other newspapers to privacy issues in relation to surveillance technologies. The article stated that:

> The nightmare scenario is that all the information held about you in different places could someday be joined together, giving whoever had access to it a total picture of your life, including many things you would rather keep to yourself ... There are those that argue that people who live good, law-abiding lives have nothing to fear from CCTV cameras, identity cards and Government plans to share their information with private companies. They would do well to remember that things can, and do, go wrong when technology is used to record the behaviour of millions of people. Privacy is a valuable thing.

In March 2002, Amanda Dowler, also known as Milly, was reported as missing.[5] *The Guardian* mentioned CCTV numerous times, but CCTV did not actually play a part in the investigation at this time (26 March 2002). In early April, CCTV footage of the girl was released (7 April 2002). In July 2002, as a suspect was arrested, *The Guardian* mentioned CCTV numerous times even though the footage recorded was of no use – it was obscured by the glare of the sun (13 July 2002). *The Times* mentioned CCTV the day after the body of Amanda Dowler was found, also in the context of the images recorded being of no use to the investigation:

> A closed-circuit television camera swiveling on top of a Bird's Eye plant 50 yards away had surveyed the area, but such was the glare from the sun that for many months detectives were unable to make any use of the footage. Earlier this month, the FBI succeeded in enhancing that film and a black sports car with a spoiler could be seen stopping alongside a figure who may or may not have been

5 Amanda Dowler (aged 13) went missing on 21 March 2002 on the way home from school. Her remains were found on 18 September 2002.

> Milly but who appeared to speak to the driver. The exchange lasted between 42 and 84 seconds, the time that the CCTV camera took to swivel back to survey the spot again. The next piece of footage showed that both the car and figure had gone. (21 September 2002)

The *Daily Mirror* in particular frequently referenced CCTV in relation to the disappearance of Amanda Dowler. CCTV was reported as potentially providing 'vital clues' and 'a vital lead' (26 March 2002; 13 September 2002; 9 November 2002). The *Daily Express* also described its potential use, reporting that:

> FBI scientists have made a potentially major breakthrough in the hunt for missing Amanda Dowler after discovering images of a mystery car on murky CCTV film … previously marred by sun glare. (13 September 2002)

A month earlier, in August 2002, Holly Chapman and Jessica Wells were also reported as missing.[6] *The Times* reported CCTV footage showing the two girls at a sports centre near Holly's home before their disappearance (9 August 2002). At this time, *The Guardian* described four potential witnesses captured on CCTV at the possible time and place of disappearance (8 August 2002; 9 August 2002). *The Guardian* also speculated that the incident would not have happened if plans to install seven CCTV cameras on the street where Holly and Jessica were last seen had not been delayed (12 August 2002).

At the same time as reporting the girls as missing and the final conclusion in the form of the arrest of Ian Huntley, there was mention in *The Guardian* (18 August 2002) of 'one of the biggest shake-ups ever of British policing' (due to calls to set up a specialist police unit similar to the FBI). Although Ian Huntley was not identified through the use of CCTV, the technology was continuously mentioned in *The Guardian* both during the investigation and after his arrest. Interestingly, at the same time as numerous mentions of CCTV in the case of Holly and Jessica (even though the technology did not play a part in the investigation or subsequent conviction of Huntley), there were a number of articles published focusing on data trails, privacy, data protection, and 'Big Brother' in relation to surveillance technologies (7 September 2002). The *Daily Mirror* and *Daily Express* reported the questioning of the effectiveness of CCTV in cutting crime, through articles referring to the release of a report by the National Association for the Care and Resettlement of Offenders (Nacro) stated that there are more effective measures that can be taken to prevent crime, such as more street lights. The *Daily Mirror* went on to include a quote from Liberty, arguing that it was time for a reassessment of the worth of CCTV 'both financial and in terms of privacy' (29 June 2002). The *Daily Express*, however, balanced the negative article with a positive about CCTV, stating that 'one of CCTV's successes was in helping to

---

6 Holly Chapman and Jessica Wells went missing on 4 August 2002 in Soham, Cambridgeshire. Their bodies were found on 17 August 2002.

convict Jon Venables and Robert Thompson for the killing of toddler James Bulger after cameras spotted them abducting him from a Merseyside shopping centre in 1993' (29 June 2002).

In January 2005, *The Times* reported Sally Geeson as missing.[7] CCTV was mentioned in terms of the police checking footage to see if she was captured on film. They did not state anything in the affirmative (*The Times* 5 January 2005; *Daily Mirror* 5 January 2005; *Daily Express* 5 January 2005). Also in January 2005, *The Guardian* reported Amy Williams (14 years old and pregnant) as missing. CCTV footage of her final movements was released (6 January 2005). Ten days later, *The Guardian* described a third person as having been arrested in connection with the murder of Amy Williams:

> The latest arrest came hours after police released new CCTV footage which showed an unidentified man walking in the opposite direction to Amy soon after she was last seen. (16 January 2005)

It did not explicitly state, but implied, that it was the CCTV footage that led to the identification of the man. The *Daily Mirror* also reported CCTV in connection with the case, saying that police were 'studying new CCTV film of her alone and fit and well' (3 January 2005). Subsequently, they also detailed the release of CCTV footage showing her last movements, alongside reporting the arrest of a man in connection with her murder. The implication that the footage led to his arrest occurred once again (6 January 2005).

In February 2005, *The Times* reported the release of a Home Office study, which cast doubts on the effectiveness of CCTV in cutting crime. It was placed on page 36 of the newspaper (25 February 2005). The *Daily Mirror* placed their report of the study on page 31 (25 February 2005). The following month, an article describing the introduction of facial recognition technology by the police was included, and stated:

> Police hope that this will make CCTV images as useful as fingerprinting and DNA evidence. (14 March 2005)

At the same time, *The Guardian* covered the introduction of car number plate recognition technology, using negative terminology, such as 'Little Brother' to describe the development. It stated that:

> Britain, which already leads the world in terms of citizen surveillance, is about to tighten the screw. (25 March 2005)

---

7 Sally Geeson went missing on 1 January 2005. Her body was discovered on 7 January 2005.

In May, *The Times* featured an article describing the extent of surveillance cameras in Britain, using terms such as 'intrusion' and 'loss of privacy' to argue strongly against the widespread nature of the technology (14 May 2005). Two months later, *The Times* published an article arguing that too much information is given out with reference to loyalty cards, the Internet, and CCTV (1 July 2005). A few days earlier, the *Daily Mirror* published an article quoting the Information Commissioner at the time, Richard Thomas, as stating that ID cards are 'an unnecessary and disproportionate intrusion of privacy'. Within this article, developments in the area of ID cards were placed alongside CCTV with facial recognition, number plate recognition and satellite tracking of vehicles, as another component of the surveillance society (28 June 2005).

Interestingly, this article appeared at a time when CCTV became a prominent feature in reports of the 7 July bombings of the London Underground. Coverage of the incident and aftermath repeatedly mentioned CCTV as playing a major role in the investigation. It was implied that if there had been more cameras the bombers could have been stopped (despite coverage also talking about the bombers as looking 'completely normal', and like 'happy hikers', making it seem unlikely that more video surveillance would have stopped them. How would they have been picked up if they weren't obvious as 'terrorists'?). I go into more detail about this juxtaposition of the bombers as 'visible' and 'invisible' in a recent paper (Kroener 2013).

Although all the newspapers I have referred to continued to mention negative aspects of CCTV during the period 2002–2005, there was simultaneously an increase in the assumption that CCTV would be of use to investigations, with it repeatedly mentioned as a first port of call for the police. Newspaper coverage of the 7 July 2005 bombings and the aftermath was generally positive about the role played by CCTV. There was little mention of the technology not having stopped the bombings occurring, even though CCTV was pushed forward as a 'crime fighting' or 'crime prevention' technology by the Government. Instead the emphasis lay on the identification of the bombers after the event and the importance of CCTV footage in this task.

The language used is extremely interesting: terms such as 'caught on camera' and 'caught by CCTV' implies some form of action. Juxtaposing this with the terms used in the event of the disappearance of Holly Wells and Jessica Chapman in 2002 shows a completely different set of terminology used. In this case, CCTV footage of the girls was described in terms such as: 'possible sightings', 'spotted', 'images of the girls'. In some ways the event of two little girls going missing frames the coverage as it can only be described as an 'awful thing to happen'. With regard to the coverage of CCTV in this case, it is placed in the role of protector, with the cameras following the girls. With regard to the 7 July bombers, the cameras follow the criminals, fulfilling the role of hunter rather than protector.

CCTV continued to be portrayed as indispensable during 2006 with reference to a number of murder investigations, even prior to assessment of the usefulness of footage, or whether footage even existed. However, at the same time there was an increase in the number of articles focusing on the negative implications of widespread CCTV and surveillance. This focus on possible negative repercussions extended into 2007, although there were also numerous articles discussing the use of CCTV in high profile crimes. During the first half of 2008, CCTV came under the spotlight in a discussion of the effectiveness of the technology, with *The Times* in particular launching an attack on the amount of money spent in relation to its impact on levels of crime.

Overall, CCTV has been portrayed in a positive light in the press; it is depicted as a technology to 'capture' and 'protect' the law-abiding public from 'fiends' and 'criminals'. An anthropomorphisation of CCTV occurs with the media portrayal of the technology as 'heroes' and 'guardian angels'. The technology was frequently reported as being 'vital' to investigations; the first port of call for the police who 'trawl' and 'scour' the footage recorded. The technology in this sense has been represented in the media as a tool to reduce risk. In its role as protector, CCTV was used to address the risk posed by terrorists after the 7 July 2005 bombings. Increased installation of CCTV was posed as a solution, by the media, to the continuing threat and risk of a repeat attack.

CCTV was also often referred to in reports of missing persons and possible abductions. In this context, CCTV was used to trace the last steps, movements, and actions of the person. Although, as has been shown, press coverage at the time of the murder of James Bulger did not focus heavily on the role of CCTV, later reports mentioned its place in this investigation as extremely important. Other media coverage did, however, thrust it into the spotlight, with the footage replayed on television during news broadcasts. The moral panic arising from this crime was surely heightened by advice from the Merseyside Police at the time:

> Until this person is caught, parents must keep hold of their children because until we know who is responsible, we cannot guarantee their safety. (*BBC News* 14 February 1993)

Later reports focusing on CCTV, and its important role in the investigation, placed it in the role of a technology that will protect the public from similar incidents – it is constructed as a risk-mitigating technology to keep children safe.

Conclusion

Studies focusing on the effectiveness of CCTV in preventing crime suggest that the technology is at best an aid to preventing crime in car parks, and littering. However, in policy and media discourse it has been constructed as an effective deterrent to criminals, and an effective crime prevention technology. Large amounts of funding were poured into the installation of CCTV systems in public space during the 1990s by a Home Office seemingly undeterred by the empirical evidence emerging during this time, which contradicted the grand claims about CCTV's effectiveness in preventing crime. Public discourse surrounding the technology during this time focused on safety and security, alongside repeated claims that to give up personal privacy would be to obtain greater security. This discourse has continued to the present day, with the media and politicians presenting CCTV as a 'friendly eye in the sky', something installed and utilised to protect the public, and implying that the public should have no concerns about increasing levels of surveillance ('if you have nothing to hide, you have nothing to fear'). In local authority discourse, CCTV is portrayed as an inclusive technology, installed to protect those who take part in civil society, and consume in town centres.

In terms of consultations at the national and local level, CCTV is portrayed as a crime prevention solution. Its uses and utility are not presented as issues for public dialogue. Consultations tend to start from the point of view that – we have these cameras, now how do we inform the public about how effective they are? Consultations seek public support, rather than debate. In this chapter I have highlighted a lack of public engagement in relation to CCTV. I provided the results of a small-scale study where a failing of communication, and a lack of public engagement prior to the start of the project, ultimately led to the failure of the project. I refer to this project as a prime example of the construction of CCTV in Britain by politicians – a techno-fix, funded by large amounts of public money, to 'fix' social problems.

Although there have been instances of coverage in newspapers highlighting privacy concerns in relation to CCTV, overall reporting has portrayed CCTV in a positive light. CCTV is constructed as a protective technology, used to catch criminals, terrorists and child abductors. In contrast to the depiction of CCTV as a 'guardian angel', those who commit crimes are described as 'fiends' and 'evil'. In both policy and media discourse, the (law-abiding) public are defined as a passive entity in need of protection.

Chapter 6
International Comparisons

Introduction

What has made the situation in Britain so different from other countries' experiences of CCTV? In this chapter, I take a look at other countries' experiences with CCTV. One of the most interesting things about public space CCTV in Britain is that it has become so much more widespread and utilised than anywhere else (with the possible exception of China in terms of absolute numbers of cameras). What are the social, cultural, legal and political experiences of other countries in the sphere of CCTV? What are the factors or experiences that have hindered the rise of CCTV in other countries? I include various EU countries in my analysis, as well as the US and Canada. I only include Western liberal democracies in this chapter. There are also extremely interesting nuances and comparisons to be made with other countries, but that is outside the scope of this book.

The Regulation of CCTV

In this section, I look at the regulation of CCTV across the European Union, Canada, and the United States. When thinking about the differences in other countries' experiences with CCTV the regulatory framework is important. Regulation on CCTV (and its enforcement) has been very different in Britain. In the first section, I provide an overview of the regulatory environment in Britain. CCTV in Britain has previously enjoyed a time of being completely unregulated. Prior to the development of the 1998 Data Protection Act, CCTV was controlled through voluntary codes of conduct, which held little effect. As we have seen, CCTV in Britain has a long history and was already utilised for a variety of purposes in public space by 1998. However, there are questions surrounding the extent to which this legislation has been enforced, and whether different countries have differing levels of enforcement. How has data protection been constructed and implemented across the European Union, Canada and the United States? What are the voices and influences that have either upheld regulation in this area, or alternatively allowed CCTV systems to be installed and run with little oversight?

The Regulation of CCTV in Britain

In this chapter, the focus is on international comparisons of CCTV across the European Union, Canada and the United States. In the next section, I provide an analysis of the regulatory environment in Britain with regard to CCTV.

In the past, CCTV enjoyed a time of being completely unregulated in Britain. Prior to the development and implementation of the 1998 Data Protection Act, video surveillance systems were 'largely controlled only by erratic, inconsistent and ill-maintained voluntary codes of conduct' (Edwards, L. 2005).

The Data Protection Act 1998 contains eight principles, which state that data must be: fairly and lawfully processed; obtained only for specified purposes; adequate, relevant and not excessive; accurate and kept up to date (where necessary); not kept for longer than necessary; and processed in accordance with the rights of data subjects (defined in the Data Protection Act). The final two principles state that appropriate technical and organisational measures should be taken to protect against unlawful or unauthorised processing of personal data, and that personal data should not be transferred to a country or territory outside the European Economic Area (unless that country or territory also has adequate safeguards in place to protect the processing of personal data). The Information Commissioner's Office states that 'most uses of CCTV will be covered by the Data Protection Act'. Any covert use of video surveillance conducted by law enforcement agencies (as well as other covert surveillance activities) is covered by the Regulation of Investigatory Powers Act (RIPA) 2000, or the Regulation of Investigatory Powers (Scotland) Act (RIPSA) 2000. The oversight of covert surveillance practices is under the remit of the Office of Surveillance Commissioners. There are also exemptions to the Data Protection Act. These include data, which is held for the purpose of preventing or detecting crime (if disclosure would prejudice that investigation), and data held for reasons of national security.

Under Article 8 of the European Convention on Human Rights Act – enacted in the 1998 Human Rights Act nationally – every individual has a right to privacy in a public place; in theory at least. This is an extremely difficult thing to prove in public – to prove and verify that your personal privacy has been compromised by a system installed legally. The definition of personal data or personal information is also blurry, both at the national and European level.

The role of Information Commissioner (and the Information Commissioner's Office) has grown out of the first position of Data Protection Registrar, set up in 1984 to oversee the new Data Protection Act. In January 2001, the Freedom of Information Act also came under the remit of the Data Protection Registrar, and the name of the office was changed to the Information Commissioner's Office (ICO). Since 2001, the ICO has come under fire for having 'no teeth'. Since 2010, the ICO has had the power to issue monetary penalties (up to £500,000 for serious breaches of the Data Protection Act). At the time of these monetary fines coming into force, Christopher Graham (the current Information Commissioner) stated that:

> I remain committed to working with voluntary, public and private bodies to help them stick to the rules and comply with the Act. But I will not hesitate to use these tough new sanctions for the most serious cases where organisations disregard the law.

Since 2011, the ICO have also been able to impose fines of up to £500,000 under the Privacy and Electronic Communications Regulations. Whether the ability to impose fines in theory translates into actually carrying this out in practice (and not just for private sector organisations but also local authorities, or national government breaches of data protection and communications laws) remains to be seen.

The Protection of Freedom Act 2012 now requires that the Government implement a code of practice and a surveillance camera commissioner to regulate the use of video surveillance systems. This Act led to the appointment of Andrew Rennison as Surveillance Camera Commissioner in September 2012. The code of practice contains guidance for those operating video surveillance systems.

The European Union

At the European level, the Data Protection Directive 95/46/EC regulates the processing of personal data. Alongside this, all member states are signatories of the European Convention on Human Rights (ECHR). Currently most national data protection is based on the following eight EU principles of data protection:

Data must be:

1. Fairly and lawfully processed
2. Processed for limited purposes
3. Adequate, relevant and not excessive
4. Accurate
5. Not kept for longer than necessary
6. Processed in accordance with individual's rights
7. Secure
8. Not transferred to countries without protection

However, this Directive currently only provides guidance, which is then implemented in national data protection regulation across the EU. Data protection law is not currently harmonised across the EU – the Directive is an attempt at harmonisation but in reality legislation at the national level overrides this EU-wide framework. However, the current proposal to reform the EU Data Protection Directive includes a proposition to transform this guiding framework into a Regulation, which is directly applicable in the Member States.

Alongside these principles of data protection, the EU upholds the protection of personal data through the Charter of Fundamental Rights of the European Union. This states that:

1. Everyone has the right to protection of personal data concerning him or her.
2. Such data must be processed fairly for specified purposes and on the basis of the consent of the person concerned or some other legitimate basis laid down by law.
3. Everyone has the right of access to data which has been collected concerning him or her, and the right to have it rectified.
4. Compliance with these rules shall be subject to control by an independent authority.

These independent authorities are typically data protection, privacy or information commissioners. The data protection principles are not self-enforcing.[1]

In terms of data protection with regard to video surveillance the Article 29 Working Party (which is made up of a representative from the data protection authority of each EU member state and provides expert advice on data protection to member states), states that:

> Data subjects have the right to exercise their freedom of movement without undergoing excessive psychological conditioning as regards their movement and conduct.

This is followed by a warning against a:

> Disproportionate application of video surveillance in public places which would allow tracking of individuals' movement and/or triggering 'alarms' based on software that automatically 'interprets' an individual's suspicious conduct without any human intervention.

The EU Data Protection Directive is currently under-going a major reform. This reform consists of a draft Regulation setting out a general EU data protection framework and a draft Directive on the processing of personal data for criminal matters. The European Parliament and Council of the EU are currently discussing these proposals, which must be approved by both prior to becoming law. On 21 October 2013, the European Parliament's leading Committee on Civil Liberties, Justice and Home Affairs (LIBE) voted to approve the reforms. MEPs and Rapporteurs of the LIBE Committee Jan-Philipp Albrecht and Dimitrios Droutsas have now entered into negotiations with the Council of the EU. Although the majority of this reform concentrates on the Internet and personal data, there are also important potential changes for the use of video surveillance particularly in

1 Which means they do not hold within them a guarantee of enforcement.

terms of 'digital traces'. The reform proposes to reinforce individuals' rights, to strengthen the EU internal market, to ensure a high level of data protection in all areas, to ensure proper enforcement of the rules and to set global data protection standards. In terms of the key changes, the EU is proposing: a 'right to be forgotten' (Article 17), which concentrates on the individual's right to decide when they no longer want their data to be processed, which will lead to the deletion of data; expiry dates with regard to the deletion of data; that consent will have to be given explicitly rather than be assumed; that individuals will have the right to refer in all cases to their home national data protection authority, even when their personal data is processed outside their home country; companies and organisations will have to notify serious data breaches (Article 32) without undue delay (and, where feasible, within 72 hours); a single set of rules on data protection, valid across the EU; increased responsibility and accountability for those processing personal data; a need for privacy protection to be included throughout the entire life cycle of a product; and finally that national data protection authorities will be strengthened so that they can better enforce the EU rules at home.

Data protection principles in the context of video surveillance are only applicable where a processing of personal data takes place. Personal data means any information relating to an identified or identifiable natural person (one who can be identified, directly or indirectly, in particular by reference to an identification number or to one or more factors specific to his physical, psychological, mental, economic, cultural or social identity). Video surveillance data has to comply with data protection principles:

> Even if the images are used within the framework of a closed circuit system, even if they are not associated with a person's particulars, even if they do not concern individuals whose faces have been filmed, though they contain other information such as, for instance, car plate numbers or PIN numbers as acquired in connection with the surveillance of automatic cash dispensers, irrespective of the media used for the processing, the technique used, the type of equipment, the features applying to the image acquisition and the communication tools used. (Article 29 Working Party)

Furthermore, data protection safeguards are in place, including obtaining and processing personal data lawfully, the proportionate deployment of a video surveillance network, and minimum intervention in terms of the use of video surveillance.

Canada

Although there is no explicit right to privacy in Canada's constitution, there are two data protection laws in operation at the federal level. These are: the Privacy Act and the Personal Information Protection and Electronic Documents Act. Further, Section 8 of the Canadian Charter of Human Rights and Freedoms provides protection against unreasonable search and seizure, thereby granting a form of privacy rights.

With reference to video surveillance, Canada's privacy guidelines stipulate adequate signage in order to obtain 'informed consent' (i.e. the public must be notified when CCTV cameras are in operation). These signs should not just indicate that video surveillance is in operation but also include information on why the cameras are there, and how members of the public can gain information on information captured. Recent research suggests that (much in the same way as in Britain) signage does not fulfil these requirements (Deisman et al. 2007).

The United States

The constitution of the United States (US) does not explicitly include the right to privacy. There is however a limited right to privacy granted by the Supreme Court, which states that an individual has a 'constitutionally protected reasonable expectation of privacy' from government surveillance (Katz vs. United States 1967). In terms of surveillance carried out by the government and private owners there is a large array of legislation centring on the use of technologies for this purpose. However, very little of this legislation applies directly to video surveillance (Gellman 2005, p.7). Even the general surveillance legislation that exists in the US lacks 'clarity, coherence, consistency, compactness, and currency … it cannot be a surprise, therefore, that the law governing video surveillance is uncertain' (ibid. p.8). Furthermore, it has been suggested that 'meaningful legal strictures on government use of public surveillance cameras in Great Britain, Canada and the United States are non-existent' (Slobogin 2002, p.233). In much the same way as the UK, the United States has developed guidelines for the operation and installation of CCTV cameras in public space, but has not set up a method of effective enforcement (ibid.).

EU/US Transatlantic Agreement

The Article 29 Working Party recently (November 2010) called for a general privacy agreement between the European Union (EU) and the United States (US). This initiative was proposed and supported by the Article 29 Working Party due to its potential for ensuring a high level of privacy protection for the individual (European Commission Press Release 19 November 2012).

Resulting from a plan by the European Commission to reach a deal with the US in relation to the exchange of personal data in criminal justice and police matters, and the protection of this personal data, the agreement attempts to set new and long-term standards in international data transfer and co-operation. This agreement also aims to reach one overarching agreement, rather than the disparate agreements in place currently: Europol (exchange of data); Extradition; Mutual Assistance; PNR (passenger name record); SWIFT (financial transactions); Container Security Initiative (CSI); Eurojust (Statewatch 2012).

Video Surveillance across Europe

Countries across Europe have vastly different levels of use of video surveillance and for various purposes. In relation to public transport, Belgium and Austria have CCTV installed on their underground networks in Brussels and Vienna. In Austria, the Wiener Linien, which manages public transport in Vienna, owns more than 1000 cameras, which are installed on the underground (Ney and Pichler 2002, p.3). The 'Verkehrsleitzentrale' (traffic management controller) owns 60 cameras. The main use of video surveillance in Austria is for traffic management and minor offences. Video surveillance is also in use to protect government and ministerial buildings, however the images are not recorded as the cameras are used for immediate security only (ibid. pp.4–5). Belgium also uses CCTV for traffic management purposes and the Brussels ring road has been equipped with cameras since 1993 (Parliamentary Assembly 2008). As far back as 1998, the European Commission of Human Rights commented that: 'Surveillance by means of video cameras, both by public and private authorities, is developing very rapidly in Belgium' (ibid.). There has also been an increase in the use of CCTV in the Czech Republic over recent years, from both private institutions and local government (Privacy International 2003).

In Denmark, the use of CCTV for surveillance purposes is generally forbidden, although exceptions to this are made for owners of certain types of property, such as petrol stations. However, if CCTV is used the owners of the property must abide by strong regulations to inform those being surveilled of the presence of cameras. These regulations are reportedly abided by in Denmark (Gras 2004). In general, the Nordic countries (of Norway, Denmark, Finland and Sweden) have resisted the use of open street CCTV systems for crime prevention purposes, however they have utilised CCTV in a variety of other ways including on public transport, in taxis and in schools (Norris 2009).

The Netherlands legislated against the use of covert video surveillance in 2004, making it unlawful to install cameras without notification (Privacy International 2007). The growth of video surveillance in the Netherlands has, however, been fairly rapid. The first public space cameras were installed in 1997 and by 2003, 80 of the 550 municipalities were using CCTV (Norris et al. 2004). Slovenia also enforces regulation of video surveillance, with their use covered in their 1999 Personal Data Protection Act and 2003 Private Protection Act (Privacy International 2007). Video surveillance systems can only be operated under license, and the public must be informed that surveillance is taking place, for what reason it has been installed and is being used, and where they can find information on data retention periods. Failure to notify the public of the presence of cameras carries the risk of fines. The last few years have seen the Information Commissioner take an active role in investigating unlawful video surveillance, leading to increased legal provisions and stricter enforcement of regulation in the installation of cameras.

In Germany, the deployment of CCTV in public space is repeatedly challenged. In 2002 there were less than 100 cameras in public areas, in comparison with the UK, which has over 40,000 in 500 cities (although these are estimates) (Hempel and Töpfer 2002). However, there is some evidence of a creeping increase in video surveillance. Since 2004 the number of cities in Germany with CCTV doubled from fifteen to thirty (Töpfer 2012). Private space video surveillance is also reportedly increasing (Goold 2004, p.24). The Czech Republic has also seen an increase in CCTV in recent years, from private institutions and local government. In the capital, Prague, it is estimated that there are two hundred cameras in the city centre, which are now linked to an automatic facial recognition system (Norris et al. 2004). Over recent years, there has also been an increase in video surveillance in Lithuania, with little or no notice given to the public of their installation. There is no legal regulation governing the use of CCTV systems in Lithuania (Privacy International 2004). Poland has recently installed their largest system of CCTV, with nineteen cameras installed in the city centre and to cover four schools, in June 2007. Italy has seen an increase in CCTV systems installed in sports premises, as well as in certain areas of its cities (Norris et al. 2004).

To date, across Europe, 21 countries have installed video surveillance systems in public space for the purpose of crime prevention (Norris 2009). These countries are: Austria, Bulgaria, Croatia, the Czech Republic, Denmark, Finland, France, Germany, Greece, Hungary, Ireland, Italy, Lithuania, Netherlands, Norway, Poland, Portugal, Spain, Sweden, Switzerland, and the UK. In the following sections I look more closely at a few examples of the use of CCTV across the European Union, Canada and the United States.

Italy

In 1981, Italy ratified the Council of Europe's Convention no. 108 for the 'Protection of Individuals with regard to Automatic Processing of Data'. However, it was not until 1996 and the passing of Act no. 675 for the 'Protection of Individuals and Other Subjects with regard to the Processing of Personal Data' that Italian data protection law was brought into line with its international obligations. Act no. 675 also incorporated Directive 95/46/EC into national data protection legislation. This Act provided a general regulatory framework for data protection, and has since been transposed by the Italian Personal Data Protection Code (2003).

Under Article 7, individuals have the right to receive confirmation of the existence or not of personal data of which he/she is subject. Furthermore (and in summary), the individual has the right to know the origin of personal data, the purposes for and conditions in which said data is to be treated, and the identity of the controller and processors, the subject or subject categories to which personal data may be communicated. The individual also has the right, wholly or in part, to oppose the treatment of data of which he/she is subject (even if pertinent to the purposes of the data collection), and to oppose the treatment of data to which he/she is subject for the purpose of commercial communications.

In the realm of video surveillance, the Italian privacy watchdog issued new regulations to protect the public in April 2010. These included the need for clear signposting for all areas under surveillance, except CCTV installed for public security purposes (e.g. the prevention of terrorism). With reference to processing personal data and video surveillance, the guidelines from the Italian Data Protection Authority state: 'image-collecting systems should be carried out in accordance not only with data protection legislation, but also with the requirements set forth in other pieces of legislation where applicable'.

Under Italian legislation the images should not be retained for longer than a few hours, and up to a maximum of 24 hours, except in the case of high-risk activities performed by the data controller (e.g. in the case of banks). Even in these high-risk cases, the period of data retention should be no longer than one week.

Germany

The first CCTV cameras used in Germany were for the purpose of traffic control and management, installed in Munich in 1958. During the following year, Hannover installed a system for traffic management, specifically for a large industrial conference and aeronautics exhibition (Weichert 1998). Hannover was the first German city to use CCTV to control 'fringe- and problem groups'. In this instance, 25 permanent CCTV cameras were installed in 1976 (ibid.).

Germany's Constitutional Court developed the 'Recht auf Informationelle Selbstbestimmung' in 1983. Essentially, this concerns the right of the individual to take control of their personal data and to determine how it is applied and to whom it is given. This development derived from the right to personal freedom described in the 'Grundgesetz' (German Basic Law), and denotes that any unauthorised collection of data on an individual goes against their civil rights and is therefore unconstitutional. However, in cases where it can be argued that the collection of data is in the public interest this collection is allowed, although it is strongly regulated and proportional (Töpfer 2003).

Video surveillance in Germany now operates within the legal framework of the European Data Protection Directive, under regulation passed in 2001. This amendment of the 'Bundesdatenschutz' (Data Protection Act), which first came into effect in 1977, incorporated the EU Directive into German law. Each state has its own further data protection act, the 'Landesdatenschutz' (Nouwt et al. 2005). The use of CCTV in public space is strictly regulated, and accompanied by ongoing public discussion of the technology, including the use of CCTV by the police (Hempel and Töpfer 2004). The installation of open-street systems continues to be minimal, although it has been rising in recent years (Helten and Fischer 2004). Those open-street systems are regulated by the state police acts, which cover the capturing of data and the length of storage time. Due to these regulations, the open-street systems in Germany have been described as monitoring 'crime hot spots' and are, in the majority, operated by the police (Hempel and Töpfer 2004). In the context of CCTV in Germany, there is an awareness of the importance of civil liberties, perhaps to a greater degree than in other European country (Spencer 1995).

France

In France, the installation of CCTV cameras must be agreed to, in advance, by an administrative authority. The registration of systems for use in public space is also compulsory. It is estimated that around 300 towns in France have installed cameras for the purpose of monitoring public space. Over recent years this number has reportedly been rising (Hempel and Töpfer 2004). France maintains a centralised system of government and despite some devolution to regional authorities, the government holds on tightly to the reins of control over surveillance technologies and the police (Spencer 1995).

The first public space CCTV cameras were installed in 1994 in a Parisian suburb. The business and financial district of Paris is continuously monitored by over 160 CCTV cameras (Nieto 1997). At present, public space CCTV can only legally be installed for specific purposes (such as the protection of property, traffic control and management, protection of public buildings) and must be authorised prior to installation. The interior of residential buildings is protected by this regulation, and the public have a right to access footage from CCTV cameras (Parliamentary Assembly 2008).

Despite this strict legislation, in 2007 the French government announced plans to dramatically increase the number of CCTV cameras (estimated to be 340,000 at the time) to over a million cameras. This announcement incorporated a promise to increase the number of video surveillance systems on the Paris metro (Norris 2009).

Spain

The use of CCTV by the police in public space is strictly regulated in Spain. In 1997, legislation concerning the use of video surveillance by the police was passed. In much the same way as in Germany, CCTV is used in Spain to monitor certain and limited locations (such as ‘crime hot spots’). Any systems in use in public space must also be registered (Hempel and Töpfer 2004). According to the 1997 legislation any footage not used for a purpose after one month must be destroyed. Individuals also have the right to access footage on which they have been recorded (Statewatch 2009).

In Spain, authorisation prior to installation of CCTV in public areas must be sought. Furthermore, any images from CCTV cameras in underground and railway stations are transmitted on monitors installed within that space, which are accessible to the public (Parliamentary Assembly 2008). There are no large networks of CCTV in Spain; instead the model is one of ‘surveillance of limited but strategic locations’ (Hempel and Töpfer 2004). CCTV cameras have also been installed in public space in the Basque region in order to help combat politically motivated vandalism by ETA supporters (Nieto 1997).[2]

In 2006, Spain published new regulation on video surveillance: Instruction 1/2006. This legislation defines images obtained by surveillance cameras in public space as personal data. This means that these images and data derived from these images are to be treated as personal data and protected as such. Under this legislation, video surveillance cameras are only to be used when other, proportionate means of surveillance are not easily available, and the collected data must be deleted within one month. Under this legislation, the processing of personal data requires the data subject's consent (although there are exceptions to this in the Spanish Private Security Law (Ley 23/992). The Spanish Data Protection Authority (AEPD) states that 'collecting the images of a person in a public place constitutes data processing'. Spain therefore recognises the right not to be filmed in public space without prior consent (Instruction 1/2006).

The United States

CCTV with facial recognition software is increasingly being piloted in public space in the US, after being used for the first time at Tampa Bay, Florida, during the 2001 Superbowl. In this instance, it was used to compare faces in the crowd with a database of images in order to spot 'potential criminals', who were then removed from the stadium (Bonner 2001). This was a joint initiative between the stadium officials, the Tampa police department and the technical and installation experts. Facial recognition software is generally considered to not be fit for purpose yet, i.e. in terms of the routine identification of an individual in a crowded area (Introna and Nissenbaum 2009).

The Lower Manhattan Security initiative was launched in 2005 in Manhattan, New York. This initiative has been compared with London's Ring of Steel. The initiative was launched as a network of CCTV cameras (alongside increased police presence and counterterrorism technologies) installed in order to increase public safety and monitor the business district (New York Police Department, Press Release, 25 February 2009). There were also plans to introduce a further 3000 cameras. As of November 2008 there were 156 cameras (*Daily News* 24 November 2008). In Washington, CCTV had originally been installed in commercial districts. However, legislation was passed in June 2006 to install cameras on street corners in residential areas (*The Washington Post* 12 October 2006).

Funding for CCTV cameras is available from the Department of Homeland Security (DHS), which had requested over $2 billion by 2005 to finance homeland security needs across the country. Some of this funding was used for setting up CCTV networks. For example, in May 2005 Chicago had a network of 2,250 cameras financed by the DHS. A further $48 million for further cameras was expected over

2 An armed Basque nationalist and separatist organisation.

the next two years. Baltimore has used DHS grants to set up a 'Watch Center' and network of cameras across the city (Electronic Privacy Information Center 2005). However, despite funding being available for the installation of CCTV, there are only a few independent evaluations of video surveillance (Monahan 2006).

Canada

Canada began installing CCTV cameras into public spaces in 1992 (Nieto 1997). Although it has been installed and used in a variety of settings, such as banks, restaurants, shops, and transport hubs, its use in public space is nowhere near as widespread as in the UK. The number of surveillance cameras installed in Canada has been limited by an active involvement on the part of the Privacy Commissioner in campaigning against their overuse (Privacy International 2007). Public opinion research conducted by marketing companies and political organisations shows high levels of support for CCTV cameras in public and private spaces. However, research conducted by academics shows significantly lower levels of support, much in the same way as in the UK (Deisman 2007). Although cameras in spaces such as transport hubs and banks have been met with seemingly little resistance in Canada, the installation of cameras to surveille public space has been widely publicly debated and in turn resisted by privacy campaigners and privacy commissioners (Bennett and Bayley 2005). One study suggests that due to this resistance 'it does not appear likely that Canada will be subject to the pervasive public monitoring in public places that is happening in Europe in the near future' (ibid.). Canadian authorities 'lack equivalent national funding for surveillance' when compared to Britain (Deisman 2007, p.10).

Most of the CCTV in Canada is privately owned and cameras monitoring public spaces, such as residential areas, are relatively low in number. However, the number of cameras monitoring public transport hubs and airports has been increasing over the last few years. This rise has been attributed to rising fear of crime and terrorism in Canada since events such as the attacks on the World Trade Centre in New York in 2001 and the London Underground bombings in 2005. Although relatively small in number in comparison to privately owned cameras, there has been an increase in the number of open-street CCTV systems in recent years. This has been attributed to a rise in fear of crime. In 2007, it was estimated that the number of cities that had installed open-street systems in Canada, was fourteen (Deisman 2007).

The Problem with Figures

It is difficult to obtain completely up to date figures on the number of video surveillance systems installed in other countries. Although the Urbaneye project (used for various references and figures in this chapter) did a very good job of documenting uses and numbers of cameras and video surveillance systems the research is now a few years old. Even in Britain it is difficult to gain accurate figures on how many cameras and video surveillance systems there actually are. The oft-cited figure of 4.2 million cameras from Norris and Armstrong (2002) is just an estimate. The BSIA did publish a report recently detailing the results of its survey of private and public sector-owned cameras (BSIA 2013). It stated that there are between four million and 5.9 million CCTV cameras in the UK (of which one in 70 is controlled by the government, and the rest are privately owned). Based on the accessible evidence, I think it is a fair judgement that Britain has the most CCTV cameras in the world (followed closely by China and the United States, in terms of absolute number of cameras). Perhaps this is not surprising, considering that we have a longer history of utilising cameras for crime prevention and other purposes. Other countries may still 'catch up'. Alternatively, it is the complex variety of social, political and economic influences that have allowed the more widespread dissemination and use of video surveillance in Britain. It is difficult to estimate what will happen in terms of other countries' uses of CCTV. However, judging by recent experiences, it is probably fair to suggest that Germany, for example, is less likely than the US to go down the same path as Britain in terms of funding, widespread use, and permanent installations of video surveillance systems in public space.

So, What has made the UK Situation so Different from Other Countries' Experiences of CCTV?

Within the surveillance studies community, there is an argument that CCTV has become ubiquitous in Britain as a result of a period of 'transformation and restructuring' under Thatcher. It is argued that this lead to heightened risk perception, social polarisation and dislocation (Norris et al. 2004). The theme of CCTV as a tool for managing risk is found elsewhere in the academic literature on CCTV (McCahill 1998). This is certainly a factor. However, other countries have also faced massive political and social upheaval yet have not seen as rapid or large a growth in public space CCTV. Perceptions of risk are also not confined to Britain.

Germany felt the after-effects of the Second World War and underwent a complete restructuring of society. From the late 1960s through to the late 1990s the country faced the threat of terrorism from the Red Army Faction (also known as the Baader-Meinhof group). With the fall of the Berlin Wall and German reunification there came yet another period of restructuring. However, this occurred alongside a strengthening of data protection laws and only limited increase in CCTV and surveillance for non-specific purposes. Germany was the first country to enact a privacy law, in 1970, concerned with the computerisation and centralisation of personal information (Klosek 2007). CCTV has been used by the German police to tackle terrorism since the 1980s. However, the general feeling has been, and remains one of, not wanting to use the cameras for more general surveillance purposes (Goold 2004). This, coupled with the decision made by the Constitutional Court to provide the individual with the right to self-determination of personal information (1983), and a continual adjusting and strengthening of data protection laws, has meant that CCTV in public space has remained restricted in Germany. Public opinion has also remained largely unchanged since the early 1980s (ibid.).

Spain has also faced massive upheaval and the 'destabilising effects of transformation and restructuring', however does not have widespread CCTV and regulates stringently. Spain did not become a liberal democracy until the late 1970s, after the death of General Franco and the end of the dictatorial regime. The structure of society and the political system was therefore entirely reinvented during the late 1970s and early 1980s, however levels of video surveillance did not increase rapidly as they did in the UK. Spain has faced numerous attacks by the Basque nationalist and separatist organisation ETA, since its founding in 1959. It is estimated that the group has been responsible for over 800 deaths since their inception (Klosek 2007). In addition, the Madrid train bombings, carried out by al-Qaeda occurred in 2004. Since the terrorist attacks on the US in 2001, Spain has strengthened its anti-terrorism laws, although a range of laws designed to combat terrorism were already in place prior to this time. For example, Article 54(1), together with 57(1) of the Law on Foreigners, allows the government to remove any foreign nationals who are believed to have participated in acts against the national security of Spain (ibid.). CCTV cameras have been installed in areas of the Basque region, but general public space video surveillance has not become prevalent.

France has also experienced numerous terrorist incidents over recent years from a variety of groups and organisations. During the 1980s those responsible for bombings in France included the pro-Iranian Lebanese Hezbollah and the Popular Front for the Liberation of Palestine and Libyan intelligence (who were also responsible in 1986 for the bombing of the Pan Am flight 103 over Lockerbie, Scotland). During 1986 and 1987 a number of bombs went off in Paris, planted by a pro-Iranian militant group (this occurred at the same time as the height of the IRA bombings in the UK). The 1990s saw a number of terrorist attacks in France, carried out by the Armed Islamic Group, the GIA (Groupe Islamique Armé). However, despite these incidents, in 1986, the French National Committee on Computer Data and Individual Freedom (CNIL) lobbied the French government to regulate the use of CCTV in public space and to design regulation to prevent potential misuse (Goold 2004, p.21). Over the next ten years the CNIL campaigned for tight controls, and in 1995 legislation was passed strictly regulating the installation of CCTV and severely limiting its use (ibid. p.22). In December 2005, anti-terrorism legislation was passed allowing greater surveillance in areas seen as high-risk (such as airports and train stations) (ibid. p.103). However, open-street CCTV systems are not included in this category.

Public opinion

Public opinion seems to have been an important factor in limiting the use of CCTV and ensuring regulation is put in place to prevent over- and misuse of the technology. France, Spain and Germany have all experienced resistance to CCTV from the public, or from an independent public body. Spain goes one step further, allowing immediate access to footage from cameras installed in underground and railway stations, thereby allowing an active public to take part not only in peer-to-peer surveillance, but to take control of information captured about them.

In Canada, although research has suggested differing results in public opinion, the Privacy Commissioner has played an important role in ensuring legislation is passed to regulate CCTV at various stages. Although the Information Commissioner in Britain has begun to play a more active role in warning of the dangers of 'sleepwalking into a surveillance society' in the last few years, the CCTV cameras have already been installed. However, even since the ICO has been provided with the means to fine those in contravention of the Data Protection Act, its involvement in actually doing so has been minimal. Britain did have a Data Protection Registrar prior to the 1998 Data Protection Act, but they did not take an active role in campaigning against CCTV cameras, or the regulation thereof.

CCTV in Britain arose at a time when public participation in technology policy decisions was fairly minimal (as discussed previously). In other European countries the wider dissemination and use of CCTV in public space occurred at a later stage when public involvement in the policy process was already occurring (although not necessarily on surveillance or security technologies). The use of CCTV was already established in Britain at this time. One survey by Hempel and Topler (2004, p.42) found that public attitudes towards CCTV were mainly positive in the five capital cities they surveyed, however attitudes differed tremendously towards open street CCTV systems. Although the majority of people supported some use of CCTV in transportation facilities and banks, the acceptance rate for the use of CCTV in public space varied considerably, with 90% of people in London seeing it as a 'good thing', as opposed to only 22% in Vienna. This is perhaps unsurprising. A longer history may well lead to a normalisation of that system or technology. Londoners are used to seeing cameras everywhere – it has become a part of everyday life. The use of public space CCTV systems in Austria has a much shorter history. Public opinion surveys during the 1990s (as has been discussed in Chapter 5) revealed a public that supposedly welcomed the use of CCTV. Although the methodologies used, the phrasing of the questions, and the lack of information about CCTV and its effectiveness, undoubtedly played a part in these results, the statistical 'evidence' was seized upon by the Home Office, politicians and the Police, keen to promote the idea of a public that felt safer due to the use of CCTV.

Legal situation

The differing legal situations, across the countries included in this chapter, have had an impact on how CCTV has been installed, and to what extent it has been used, for what purposes, and so on. There is no constitution in Britain, and therefore no right to privacy written into the constitution. Furthermore, Britain does not have any specific legislation regulating the installation and operation of CCTV cameras. Even under the Data Protection Act, what is deemed to be collection of 'personal data' is debatable. The UK Information Commissioner has ruled that if CCTV cameras are not able to focus in on a particular individual, there is no contravention of the Data Protection Act and no need notify the use of the system to the Information Commissioner's Office (Edwards 2005). The UK does have a Code of Practice for these systems, but it is largely ignored (Moran 2005). Some more sophisticated systems are still subject to data protection controls, for example if footage is taken for the purpose of learning about the activities of a specific individual, rather than footage of general scenes being recorded, data protection law is applicable. However, cameras used for crime control and crime prevention purposes are exempt from data protection legislation as their purpose is concerned with national security and crime prevention. In addition, courts in the UK are willing to accept evidence from illegally installed CCTV cameras. It could be argued that this does not promote good practice (Edwards 2005).

The European countries discussed in this chapter either have specific regulation governing video surveillance (France, Spain and Italy) or stringent data protection laws (Germany). In contrast to Britain, the legal situation surrounding the installation of public space CCTV systems was not left wide open. As mentioned previously, Canada does not have specific regulation governing the use of CCTV systems, but public involvement in the policy debate and an active privacy commissioner has meant that widespread dissemination of the technology in public space has not occurred. As I have illustrated, the growth of public space systems has been greatest in the United States. It seems that the lack of legal regulation governing CCTV is leading to a similar situation to that in Britain – the growth of systems without enforceable guidelines. Furthermore, CCTV cameras in the US and Britain are installed without requiring prior permission obtained from a controlling authority. Any analysis of the actual need for CCTV in that area of public space has therefore become minimal. Rather, the CCTV system is installed and at some point a post hoc evaluation of its effectiveness is carried out.

Funding

The amount of funding provided by central government for the installation of CCTV systems is certainly a major factor in the relative growth of video surveillance internationally. No other country has committed the strength of resources to CCTV that Britain has. During the various Challenge Competitions ignited by central government, it was not just the amount of money made available to local authorities that prompted a greater interest in installing CCTV systems in town centres in Britain, but the competitive element to the distribution of funds that sparked a more aggressive want for video surveillance in certain areas.

A further similarity between Britain and the US is the availability of funding for CCTV systems. As detailed previously, the US Department of Homeland Security made increasing levels of funds available over recent years for video surveillance initiatives. As we have seen, a similar situation occurred in Britain with a number of grants made available to local councils to install CCTV systems. The funding for CCTV in Britain has not been political party-specific, commencing under a Conservative government, but continued under New Labour.

Other surveillance methods
The majority of countries in the European Union have a form of identity card. Public acceptance seems to come from a long history of identity card systems and opposition only increases when changes, such as the proposed inclusion of religious denomination in Greece, occur. Neither the United States nor Canada has an identity card in operation. In both countries the topic is publicly debated and various consultations and proposals have been put forward in recent years. Despite the levels of public acceptance of one method of surveillance – the identity card – no European country has the levels of dissemination of CCTV cameras of the UK. Since events such as the terrorist attacks in the United States on the 11 September 2001 took place, many countries have redesigned and strengthened their anti-terrorism laws. For example, Spain and France implemented a system to allow the deportation of individuals, described as 'radical Islamists', seen to be exhorting others to commit acts of terrorism. The revoking of citizenship can also occur in some cases (Savitch 2008, p.145). The UK government argues that CCTV is used as a tool to combat terrorism. However, despite redesigning its anti-terrorism laws, no other country has allowed the spread of video surveillance to the extent Britain has for the purpose of combating terrorism.

Conclusion

In this chapter we have seen that CCTV in Britain enjoyed a time of being completely unregulated. The Data Protection Act was enacted in 1998, stipulating a set of principles to govern the use of personal data. There is also EU-level regulation in place to regulate the processing of personal data. However, the principles contained in data protection legislation are open to interpretation at the national level. Regulation is constructed and defined in the national context. We have seen, for example, that consent can mean a vastly different thing from one country to the next.

Funding has also played a major part in the amount of CCTV installed in the various countries analysed in this chapter. No other country has seen such vast amounts of money poured into installing CCTV systems in public space as Britain. The competitive nature of the funding allocation added a further dimension to the dissemination and use of CCTV by local authorities.

Britain has an Information Commissioner, with the ability, in theory, to implement fines for contraventions to the Data Protection Act. However, the extent to which these fines are actually applied in practice varies. There is no privacy commissioner in Britain. In those countries with either a privacy commissioner, or stringent privacy laws, the dissemination of CCTV has been more cautious. Public opinion has also played a role in how widespread the use of CCTV in Britain is. There has been relatively little public resistance to CCTV. Coupled with a lack of resistance from a public body (such as the work of a privacy commissioner), the reason why CCTV is installed so routinely in public space in Britain is barely questioned. Political rhetoric surrounding the technology is therefore allowed free reign, and politicians and the police are able to use the results of public opinion surveys (with questionable phrasing) to justify the use of a technology that, they suggest, 'makes the public feel safer'. In the next chapter, I continue with a call for greater, and different, public engagement in the area of CCTV, and surveillance more generally.

Conclusion

> [Technology] absorbs meaning through its use and what it symbolises to others.
> (Boradker 2010)

Why has Britain become so camera-surveilled? What were the factors involved in its development and subsequent use as a technology to protect people from crime? How has an ostensibly mundane technology become such an important tool for politicians and the police? And how did the widespread deployment of a surveillance technology occur seemingly under the radar of the public?

This book uses CCTV as an example but the questions raised also hold weight with all sorts of other technologies and technological systems. How are the technologies that we use, or come into contact with every day, constructed and defined? How did they become what they are? Science and Technology Studies (STS) tells us that technologies do not develop outside of society. They are a part of it – influenced, shaped, and designed by it. Their meaning is 'flexible' (Bijker 1995). As they are applied they are also constructed. Their properties are contingent, rather than fixed.

Looking at the history of CCTV through an STS lens allows us to see the development of the technology and its present day uses as something that was not pre-determined. Implicit in theorising about technology in this way is an underlying message that the properties of the technology are not concrete. Nor are its uses.

The focus for Chapter 1 was surveillance. We have seen that an increase in the use of information and communication technologies, and surveillance technologies have led to arguments from a variety of commentators that we are now living in a surveillance society. There has undoubtedly been an increase in surveillance technologies and practices in Britain in recent years. Data is collected, processed and stored from a number of technologies, activities, and sources. The management of risk, and the management of criminal behaviour, has become tied in with a variety of strategies and technologies. In Chapter 2 we have seen the political, social, economic and cultural changes that Britain has undergone since the end of the Second World War. We have seen changes in the social sphere in the context of social divisions, housing and rising crime rates; and a divide between the public and the Police. We have also seen the politicisation of science and technology, and the existence of a belief in a techno-fix – that social problems can be fixed via technological means. In Chapter 3, we have seen how the problem of rising crime rates started to creep onto the political agenda.

In Chapters 3 and 4 we have seen that CCTV could have been otherwise. Its first uses were in the areas of transport, education, and health and safety. At its inception, CCTV was used as a transmission technology. Under a changing political climate and evolving criminal justice context, CCTV began to be utilised as a crime prevention technology and for traffic control purposes. We have seen that changes in policing meant that an alternative solution was needed in order to tackle rising crime rates. CCTV began to be defined politically as a tool for crime prevention. In Chapters 4 and 5 we have seen that a variety of influences and actors began to construct CCTV as a technology to provide safety and security. The political climate began to shape CCTV as a crime prevention technology. The discourse surrounding CCTV during this time promoted CCTV as a protective, non-intrusive, extra eye. Civil liberties arguments were discounted as unimportant in relation to the benefits of the installation of CCTV systems. This discourse has become cemented over the last two decades with policy and media discourse supporting a myth of effectiveness. Empirical investigations focusing on the results of CCTV have at best found it to be useful in terms of reducing crime in car parks and instances of littering. Studies that have found a reduction in overall crime statistics have been dismissed as 'post hoc shoestring efforts by the untrained and self-interested practitioner' (Pawson and Tilley 1994, p.294) or 'wholly unreliable' (Short and Ditton 1995, p.10). Despite the evidence to the contrary, the popular discourse surrounding CCTV has been that it 'works'. CCTV has become politicised. It has come to represent a solution for politicians – a symbolic gesture to show that they are 'doing something' about crime. In turn, CCTV systems also shape and define society. Society has become constructed and defined as based on surveillance practices, categories of individuals, and subjects of surveillance. It has also become defined as a society that is willing to accept and support CCTV in order to gain security. Furthermore, in contrast to the politicisation of the technology by politicians and the police, CCTV has also become depoliticised as a civil liberties problem.

In Chapter 6 we have seen that other countries have had vastly different experiences with CCTV. The legal frameworks operating in the different countries analysed have clearly had an influence over where and how CCTV is installed and for what purposes it is used. It is not just legislation and regulation that have had an impact though. Countries without specific regulation governing the use of CCTV have still seen more limited use due to having an active privacy commissioner (for instance, Canada). In terms of funding, no other country has seen the levels of funding provided for the installation of video surveillance systems in public space that Britain has. Public opinion has also played a role. CCTV in Britain has a long history. At the time of its inception it inhabited a very different role to that which it has now (it was seen to be more or less an extension of broadcasting, with the exception of limited police experiments around traffic control and public space surveillance). CCTV gradually started to be used as a tool for surveillance during the 1980s (although various experiments were conducted by the police prior to this time), when public engagement around science and technology was focused on environmentalism and renewable energy. Its dissemination slipped under the radar of social protest movements.

The Discourse Surrounding CCTV

What has CCTV come to represent in twenty-first-century Britain? When and why did a mundane technology used for a variety of purposes become constructed and defined as a surveillance technology? What were the voices and influences that played a part in this construction? How has the myth of the effectiveness of CCTV been perpetuated in public discourse?

Throughout this book, we have seen that the public discourse surrounding CCTV emerged in the 1960s with experiments by the police in a variety of locations. A now familiar discourse began to emerge during this time suggesting that CCTV was no different to a police officer standing on a street corner. There were also connotations of surveillance capability in other public discourse. CCTV was reported as an 'extra eye' on transport vehicles, in playgrounds, and for health and safety purposes. However, during this time CCTV also inhabited a variety of other uses (transmission, education, as a tool to practice TV broadcasts).

During the 1970s the problem of crime crept onto the political agenda. Politicians, keen to be seen as part of the solution to tackling rising crime rates, began to develop a discourse focused on punishment of the 'criminal', and protection of the 'victim'. This discourse has developed during the 1980s and to the present day. The idea of condemnation of deviance is a strong discourse that divides society into those who are law-abiding, and those who aren't. The use of CCTV fits in easily with this discourse. It is used as a highly divisive technology. It is constructed as a technology that is there for the protection of those who do not break the law, and utilised to 'catch' those who do.

We are told that CCTV is there to protect the public and to prevent crime, and that the technology should be accepted under these terms. Potential concerns over an erosion of civil liberties are dismissed in light of apparent results in terms of crime prevention and detection. The use of CCTV has become synonymous in political discourse with enhanced safety and security. The public have been constructed as a group who welcome the use of CCTV. We are told that it makes us 'feel safer'.

Public Resistance

Miller and Woodward (2007, p.337) argue that ubiquity generates invisibility. Has CCTV become so widespread that it has simply become invisible? There are cases of public resistance to certain types of video surveillance, for example, speed cameras. This is interesting. Rather than being categorised as 'law-abiding', the public are suddenly part of 'the other' in society – the people that the cameras are out to catch. Heidegger has argued that we only notice technologies when they stop working in the way that we anticipate (Brown 2004). In terms of speed cameras, video surveillance suddenly inhabits a different form and different context. They no longer 'work' for the purposes anticipated.

Resistance to CCTV in general is complex and multi-faceted. Although public debate has been minimal there are instances of resistance to the installation of systems. One of the most prominent examples of public pressure to remove CCTV cameras occurred in Birmingham in 2010/2011. 'Project Champion' saw the installation of 200 cameras, funded with £3 million of government money earmarked for 'fighting terrorism' installed at various locations throughout the areas of Sparkhill, Ward End and Washwood Heath. Predominantly Muslim areas, the installation of these cameras sparked a protest from members of the public, civil liberties groups, and other campaigners, eventually culminating in the cameras being taken down. Members of the public reported feeling angry about a lack of public consultation prior to the cameras being installed (BBC News 25 October 2010).

There are also examples of other forms of resistance in Britain. We have seen not only directed public pressure towards a CCTV system (as was the case in Birmingham) but also opposition from the House of Lords, activity by civil liberties groups, work by graffiti artists through other art forms, and literature (Martin et al. 2009). In recent years, civil liberties groups have played an important role in campaigning against the increasing use of surveillance technologies. However, it has become apparent that the majority of the public have accepted CCTV and its use in public space (Lyon et al. 2012, p.4).

Public Engagement

> One thing about which fish know exactly nothing is water, since they have no anti-environment which would enable them to perceive the element they live in. (McLuhan 1968)

As I suggested earlier, the argument that we are already living in a surveillance society might imply that there is little, or nothing, left to resist. I argue for an open dialogue about CCTV (and surveillance), based not only on privacy – which can seem abstract and is often dismissed through the rather empty statement: 'If you have nothing to hide you have nothing to fear'. These sorts of discussion shut down any meaningful dialogue. Instead we have an argument between two groups of people: those who care about their civil liberties and those who don't. It is time to open up the debate and challenge the discourse.

In a liberal democracy, providing an arena for the public scrutiny of surveillance technologies prompts discourse and debate on developments through which the public are directly affected. As we have seen, CCTV is often portrayed as an anti-terror 'silver bullet' (Fay 1998), or a techno-fix for problems of crime.

Amongst surveillance technologies, CCTV is of paramount importance: it is the most ubiquitous surveillance technology in Britain (which potentially leads the world in deployments of CCTV cameras). The construction of the public in terms of policy and media discourse remains firmly situated in a deficit model of public understanding. Despite government policy shifts in recent years from science communication to public engagement suited to 'contentious contemporary social issues' (Gregory and Lock 2008), within the policy and media discourse on CCTV over the last twenty years, the public have been constructed as a passive entity in need of protection. This stands in contrast to a general trend in policy in which the public are viewed as active participants in democratic decisions about science and technology (Gregory 2001).

That is not to say that there are not any problems with public engagement on science and technology more generally. Wilsdon and Willis (2004), for example, argue that the voices of more 'diverse publics' need to be heard in discussions on science and technology, and that 'only by opening up innovation processes at an early stage can we ensure that science contributes to the common good'. This is, of course, easier said than done. However, there have been important moves towards experiments in 'processes of dialogue and participation'. This has not happened in the area of CCTV (or in relation to other surveillance technologies), which is interesting as CCTV is often installed by authorities that have public democratic oversight (local councils).

Engaging people's thoughts, expectations, concerns, interests and reactions to everyday surveillance is vital in terms of stimulating public debate and developing accountability and participation in relation to open government. This is of crucial importance given the ubiquitous nature of CCTV, the huge public expenditure on its development and usage, and the impact of surveillance on the everyday lives of the public. Engagement in this area cannot simply take the form of a public opinion survey through which members of the public have the limited opportunity to answer questions that are deemed relevant by policy-makers, academics, or journalists, which often show considerable bias towards the use of CCTV.

Rather than a consultation designed to ensure public support for 'the security measures that are in place', a full and informed public debate on these issues should follow a model of 'upstream engagement' (Wilsdon and Willis 2004). This model emphasises a 'substantive motivation' with the eventual outcome one of influencing policy decisions, improving decision-making processes, and to create more 'socially robust scientific and technological solutions' (ibid. p.39). So, rather than the public taking part in a consultation process designed to justify decisions already made (as one could argue is the case with the recent Protection of Freedoms Bill) or to verify and support particular interests, an upstream engagement process is based on the public 'work[ing] actively to shape decisions, rather than having their views canvassed by other actors to inform the decisions that are then taken' (ibid.).

Recently, there have been a number of deliberative engagement processes (some examples of these can be seen on the ScienceWise-ERC website). These processes are now becoming more prevalent in Britain. There is not one ideal methodology or approach to practicing meaningful public engagement but what needs to be remembered is that the public must be involved in not only expressing opinions on a given issue (with questions framed by 'experts' in the field) but rather that they also be given the opportunity to shape the questions that are asked and that they consider to be important (Stirling 2008).

Public Engagement in the 'Surveillance Society'

As I have mentioned previously, it is worth thinking about whether statements that we live in a surveillance society simply reinforce an already strong discourse around CCTV. These sorts of statements imply that there is little scope for change. I believe that a move towards public engagement in relation to CCTV, rather than a focus on the surveillance society would be beneficial.

There are three possible arguments for public dialogue. The first is that dialogue leads to more democratic decisions. The second is that it can lead to better decisions, due to a wide range of expertise being drawn in. The third is that dialogue leads to more socially robust decisions. Although all three arguments have validity, I believe the third argument to be the most important with regard to CCTV.

There have been a number of consultations at the national level on security technologies in recent years, although those with a specific focus on CCTV have been minimal. Those that have focused specifically on CCTV have sought public views to reinforce and justify already existing security measures, rather than opening up the consultation process in a way to promote public dialogue about the wider issues related to video surveillance.

There are also examples of consultations at the local level focusing specifically on CCTV. As has been seen at the national level, these local consultations seek to reinforce notions of the effectiveness of CCTV in preventing crime. Rather than promoting any meaningful dialogue, involving a range of experts and lay persons, they tend to concentrate on 'informing' the public about the benefits of using CCTV in public space. Lay advisory panels can therefore become a simple portal for the dissemination of established discourse surrounding the effectiveness of CCTV.

There are, of course, a number of problems with attempting to undertake public engagement. There are questions of: who should facilitate discussions? Should this be undertaken at the national or local level? How long are results socially valid? These are all important questions and there is no right answer. We need to move towards at least discussing these issues in the area of CCTV and surveillance. I strongly believe that we need a privacy commissioner in Britain. This could allow for a set of guidelines to be drawn up outlining provisions for determining the necessity of video surveillance as a last resort (Office of the Privacy Commissioner, Canada), rather than allowing the continuing routine installation of CCTV as an unquestioned default solution. Furthermore, this could also allow a public forum for the dissemination of research results from academics and other researchers working in the area of surveillance and society issues. Currently, surprisingly little academic research makes it out into the public domain or policy arena in relation to CCTV or surveillance. When it does, it tends to reach the public in a diluted form, used to 'justify preconceived policies under institutional pressures' (Lett et al. 2010).

However, as I have stated previously, privacy is not the only issue with regard to CCTV and surveillance more generally. Debate and discussion in this area needs to move beyond an argument over an eradication of civil liberties, to encompass wider issues of the use of public resources and the potential for other solutions to crime to be developed at the local and national level. The current lack of oversight in the installation of CCTV systems in turn results in a lack of accountability with regard to its uses and results. There is an urgent need for greater public participation in the policy process on CCTV, and a need for those installing and using CCTV systems to be held to account. Current developments in CCTV are moving towards 'smart systems' with built in algorithms for 'detecting' a range of behaviours. The implications of these systems need to be debated and evaluated now. The proliferation of surveillance technologies may give reason to fear that it is too late to change things, and that our entry into a surveillance society is complete. I believe that it is not too late. There are still questions to be answered. It is time to contest the political discourse surrounding CCTV and to open up discussions to a wider, engaged and active public.

References

Agar, J. (2005) 'Identity Cards in Britain: Past experience and policy implications' *History and Policy* http://www.historyandpolicy.org/papers/policy-paper-33.html

Ainley, R. (1998) 'Watching the Detectors: Control and the Panopticon' in R. Ainley (ed.) *New Frontiers of Space, Bodies and Gender* Routledge; London pp.88–100

Ainley, R. (2001) 'Keeping an Eye on Them' in S. Munt (ed.) *Technospaces: Inside the New Media* Continuum: London p.86

Albrechtslund, A. (2008) 'Online social networking as participatory surveillance' *First Monday*: http://firstmonday.org/ojs/index.php/fm/article/view/2142/1949#p4

Alderson, J. (1978) 'Concepts of Prevention Policing' Peel Press: UK: https://www.ncjrs.gov/App/Publications/abstract.aspx?ID=61855

Altheide, D. (2002) *Creating Fear: News and the Construction of Crisis* Aldine de Gruyter: New York

Anderson, P. (2012) 'Fighting 'Terrorism' or Repressing Democracy? Britain's System of Mass Surveillance' Global Research: http://www.globalresearch.ca/fighting-terrorism-or-repressing-democracy-britains-system-of-mass-surveillance/5311802?print=1

Archer, J.M. (1993) *Sovereignty and Intelligence: Spying and court culture in the English Renaissance* Stanford University Press: Stanford

Armstrong, G. (2003) *Football Hooligans* Berg: Oxford

Armstrong, G. and Giulianotti, R. (1998) 'From another angle: police surveillance and football supporters' in C. Norris, J. Moran and G. Armstrong (eds) *Surveillance, Closed Circuit Television and Social Control* Ashgate: Aldershot

Association of Chief Police Officers (ACPO) *Police Review* 11 August 1956

Association of Chief Police Officers (ACPO) *Police Review* 7 May 1957

Association of Chief Police Officers (ACPO) *Police Review* 6 July 1957

Association of Chief Police Officers (ACPO) No.6 District Conference minutes (1960) 'Closed Circuit Television Use by the Police' (Open University archive)

Association of Chief Police Officers (ACPO) (2004) *E.C.H.R. Data Protection & RIPA Guidance Relating to the Police Use of A.N.P.R.* ACPO National ANPR User Group: London

Association of Chief Police Officers (ACPO) ANPR Steering Group (2005) 'ANPR strategy for the police service 2005–8: Denying Criminals the Use of the Road' ACPO: London

Bailey, V. (1981) *Policing and Punishment in Nineteenth Century Britain* Croom Helm: London

Ball, K. and Murakami Wood, D. (2006) (eds) 'A Report on the Surveillance Society for the Information Commissioner' http://www.ico.org.uk/upload/documents/library/data_protection/practical_application/surveillance_society_summary_06.pdf

Bannister, J. Fyfe, N.R. and Kearns, A. (1998) 'Closed Circuit Television and the City' in C. Norris J. Moran and G. Armstrong (eds) *Surveillance, Closed Circuit Television and Social Control* Ashgate: Aldershot pp.21–39

Barnard-Wills, D. (2011) 'UK News Media Discourses of Surveillance' in *The Sociological Quarterly* 52(4) pp.548–67

Barnett, H. (2011) *Constitutional and Administrative Law* (9th edn) Routledge: London

BBC News 17 February 1993 'Missing Two-year-old found Dead'

BBC News 18 December 2003 'Ring of Steel Widened'

BBC News 16 August 2004 'Watchdog's Big Brother UK Warning'

BBC News 17 September 2006 'Town Trials Talking CCTV Cameras'

BBC News 4 April 2007 'Talking CCTV Scolds Offenders'

BBC News 28 April 2010 'Do CCTV and the DNA Database Make Us Safer?'

BBC News 25 October 2010 'Police Chief Wants Birmingham Spy Cameras Removed'

BBC News 12 August 2013 'Surveillance Camera Code of Practice Comes into Force'

Beck, U. (1992) *Risk Society: Towards a new modernity* Sage: London

Bennett, C.J. and Bayley, R.M. (2005) 'Video Surveillance and Privacy Protection Law in Canada' in S. Nouwt, B.R. de Vries and C. Prins (eds) *Reasonable Expectations of Privacy: Eleven Country Reports on Camera Surveillance and Workplace Privacy* TMC Asser Press: The Hague

Big Brother Watch (2012) 'London Met Police Spends 4m a Year Watching CCTV'

Bijker, W. (1995) *Of Bicycles, Bakelites, and Bulbs: Toward a theory of sociotechnical change* MIT Press: Cambridge

Bimber, B. (1994) 'Three Faces of Technological Determinism' in M.R. Smith and L. Marx (eds) *Does Technology Drive History?* MIT Press: Cambridge pp.79–98

Black, J. (2004) *Britain since the Seventies: Politics and Society in the Consumer Age* Reaktion Books: London

Bloomfield, B., Boanas, G. and Samuel, R. (1987) (eds) *The Enemy Within: Pit villages and the miners' strike of 1984–5* Routledge: London

Bonner, J. (2001) 'Looking for Faces in the Superbowl Crowd': http://securitysolutions.com/mag/security_looking_faces_super/

Boradker, P. (2010) *Designing Things: A critical introduction to the culture of objects* Berg: Oxford

Bradley, D. et al. (1986) *Managing the Police: Law, Order and Democracy* Harvester Wheatsheaf: Brighton

Brake, M. and Hale, C. (1992) *Public Order and Private Lives: The politics of law and order* Routledge: London

Brey, P. (2005) 'Artifacts as Social Agents' in Harbers, H. (ed.) *Inside the Politics of Technology: Agency and Normativity in the Production of Technology and Society* Amsterdam University Press: Amsterdam pp.61–84

Bright, J. (2011) 'Building Biometrics: Knowledge construction in the democratic control of surveillance technology' *Surveillance and Society* 9(1/2) pp.233–47

Brin, D. (1998) *The Transparent Society* Perseus Books: Massachusetts

Brown, B. (1995) 'CCTV in Town Centres: Three case studies' Crime Detection and Prevention Series, Paper 68 Home Office: London

Brown, B. (2004) (ed.) *Things* University of Chicago Press: Chicago

Browne, J. 30 September 2012 'Regulation of CCTV and other surveillance camera technology' Written statement to Parliament

Bulos, M. (1995) 'Towards a Safer Sutton? Impact of closed circuit television on Sutton town centre' London Borough of Sutton

Bulos, M. and Sarno, C. (1994) 'Closed Circuit Television and Local Authority Initiatives: The first national survey' Research Monograph, South Bank University

Burgot Rentals Ltd., correspondence, 4 May 1959 (Open University archive)

Canadian Charter of Rights and Freedoms: http://laws.justice.gc.ca/en/charter/

Castells, M. (1996) *The Rise of the Network Society* Blackwell Publishers: Oxford and Malden, MA

CCTV User Group (2010) 'An Independent Public Opinion Survey on the Use and Value of CCTV in Public Areas'

Chapman, J. (2005) *Comparative Media History, an Introduction: 1789 to the present* Polity Press: Cambridge

Chibnall S. (1977) *Law-and-Order News: An analysis of crime reporting in the British press* Tavistock Press: London

Chief Constable of Kent (January 1964) Central Conference of Chief Constables, Police Communications Committee, 'Closed Circuit Television' (Open University archive)

Childs, D. (1995) *Britain since 1939: Progress and Decline* Macmillan: London

Coaffee, J. (2004) 'Recasting the "Ring of Steel": Designing out terrorism in the City of London' in Graham, S. (ed.) *Cities, War and Terrorism: Towards an Urban Geopolitics* Blackwell Publishing Ltd.

Cohen, S. (1973) *Folk Devils and Moral Panics* Paladin: St. Albans

Cohen, S. (1985) *Visions of Social Control* Polity Press: Cambridge

Coleman, R. and Sim, J. (2000) '"You'll Never Walk Alone": CCTV Surveillance, order and Neo-liberal rule in Liverpool City Centre' *British Journal of Sociology* 51(4) pp.623–39

Crawford, A. (1998) *Crime Prevention and Community Safety* Longman: Harlow

Crime and Disorder Act 1998: http://www.opsi.gov.uk/acts/acts1998/ukpga_19980037_en_1

Daily Express 3 August 1995 'New Tactics'

Daily Express 9 November 1995 'Big Cash Boost for Crime War'
Daily Express 15 September 2000 'Bosses Spying on Staff'
Daily Express 5 December 2000 'Caught on Camera, Little Damilola's Final Living Hour'
Daily Express 4 April 2002 'There's No Hiding Place from This Army of Spies'
Daily Express 29 June 2002 'CCTV Does Not Stop Crime'
Daily Express 12 August 2002 'Jessica and Holly'
Daily Express 13 September 2002 'Milly: FBI find clue'
Daily Express 5 January 2005 'Cry for Help by Text'
Daily Mail 6 August 2002 'How Did They Just Vanish?'
Daily Mail 27 September 2013 'The Man Who Never Forgets the Face: How Scotland Yard's elite squad of 200 'super recognisers' can spot a suspect in a crowd'
Daily Mirror 23 January 1958 'Votes and Viewers'
Daily Mirror 16 November 1958 'Just Look at Me – on TV'
Daily Mirror 10 April 1995 'You've Been Framed Mr Howard'
Daily Mirror 6 March 1998 'Drunken Thug Teachers Jailed for Battering Cop'
Daily Mirror 26 August 1998 'Gang in Van Raid Bungle'
Daily Mirror 5 February 2000 'MP's Niece Murdered at Giant's Causeway'
Daily Mirror 24 May 2000 'Has Sara Murderer Struck Second Time?'
Daily Mirror 10 June 2000 'Cops Quiz Children over Rail Horror'
Daily Mirror 1 July 2000 'Lair of Hatred'
Daily Mirror 1 July 2000 'Nothing Excuses this Evil'
Daily Mirror 19 August 2000 'Rapist in Attack on Aids Girl'
Daily Mirror 1 December 2000 'Damilola on CCTV'
Daily Mirror 26 March 2002 'I Spotted Milly'
Daily Mirror 29 June 2002 'Bigger Brother's watching … But It's Just Not Working'
Daily Mirror 13 September 2002 'Is this Milly?'
Daily Mirror 9 November 2002 'Mazda Driver Link to Milly Ruled Out'
Daily Mirror 3 January 2005 'Amy Quiz: Cops get more time'
Daily Mirror 5 January 2005 'Missing Sally's Desperate Text Message to Pal'
Daily Mirror 6 January 2005 'The Last Pictures'
Daily Mirror 25 February 2005 'CCTV Does Not Cut Crime'
Daily Mirror 28 June 2005 'An Unnecessary and Disproportionate Intrusion of Privacy'
Daily News 24 November 2008 'Lower Manhattan Security Initiative Up and Running, Safe from Budget Cuts'
Dandeker, C. (1990) *Surveillance, Power and Modernity: Bureaucracy and discipline from 1700 to the present day* Polity Press: Cambridge
Data Protection Act 1998 (DPA)
Data Protection Directive 95/46/EC
Davies, M., Croall, H. and Tyrer, J. (2005) *Criminal Justice: An introduction to the criminal justice system in England and Wales* (3rd edn) Longman: Harlow and London

Davies, S. (1996) *Big Brother: Britain's Web of Surveillance and the New Technological Order* Pan Books: London

Davies, S. (1996) 'The Case Against: CCTV should not be introduced' *International Journal of Risk, Security and Crime Prevention* 1(4)

Davis, M. (1990) *The City of Quartz: Excavating the Future in Los Angeles* Vintage: New York

Deisman, W. et al. (2007) 'A Report on Camera Surveillance in Canada' Surveillance Camera Awareness Network: http://www.sscqueens.org/sites/default/files/SCAN_Report_Phase1_Final_Jan_30_2009.pdf

Dewey, J. (1927) *The Public and its Problems* George Allen & Unwin: London

Dijk, J.J.M. van (1990) 'Crime Prevention Policy: Current State and Prospects' in G. Kaiser and H.J. Albrecht (eds) *Crime and Criminal Policy in Europe: Proceedings of the II European Colloquium* Max-Planck-Institute

Ditton, J. and Short, E. (1999), 'Yes, It Works, No, It Doesn't: Comparing the Effects of Open CCTV in Two Adjacent Scottish Town Centres', in Painter, K. and Tilley, N. (eds) *Crime Prevention Studies* 10 pp.201–24

Douglas, M. and Wildavsky, A. (1982) *Risk and Culture: An essay on the selection of technological and environmental dangers* University of California Press: Los Angeles

Downes, D. and Morgan, R. 'No Turning Back: The politics of law and order into the millenium' in Maguire, M. et al. (2007) *The Oxford Handbook of Criminology* Oxford University Press: Oxford pp.201–40

Dunbar, I. and Langdon, A. (1998) *Tough Justice: Sentencing and Penal Policies in the 1990s* Blackstone Press Ltd: London

Eagleton, T. (1976) *Criticism and Ideology: A Study in Marxist literary theory* Verso: London

East, R. and Thomas, P. (1985) 'Road-blocks: The experience in Wales' in B. Fine and R. Millar *Policing the Miners' Strike* Lawrence & Wishart: London

EDRI-gram September 2010 'European Commission's Strategy for Data Protection Directive': http://www.edri.org/edrigram/number8.18/ec-strategy-data-protection-directive

Edwards, C.J. (2005) *Changing Policing Theories for 21st Century Societies* (2nd edn) Federation: Annandale

Edwards, L. (2005) 'Switching Off the Surveillance Society: Legal regulation of CCTV in the UK' in S. Nouwt, B.R. de Vries and C. Prins (eds) *Reasonable Expectations of Privacy: Eleven Country Reports on Camera Surveillance and Workplace Privacy* TMC Asser Press: The Hague

Electronic Privacy Information Center (May 2005) 'More Cities Deploy Camera Surveillance Systems with Federal Grant Money': http://epic.org/privacy/surveillance/spotlight/0505/

Ellul, J. (1962) 'The Technological Order' *Technology and Culture* 3 pp.394–421

Emmerson, A. (July 2000) 'Behind Big Brother' *Electronics World*

Emsley, C. (1995) 'Preventive Policing: The path to the present' in J.P. Brodeur (ed.) *Comparisons in Policing: An international perspective* Avebury: Aldershot

Entman, R.M. (1993) 'Framing: Towards Clarification of a Fractured Paradigm' *Journal of Communication* 43 pp.51–8

European Commission Press Release (Brussels) 19 November 2010 'Article 20 Data Protection Working Party'

European Convention on Human Rights Act 1953

Fay, S.J. (1998) 'Tough on Crime, Tough on Civil Liberties: Some negative aspects of Britain's wholesale adoption of CCTV surveillance during the 1990s' *International Review of Law, Computers and Technology* 12 pp.315–47

Fennelly, L.J. (2003) *Effective Physical Security* (3rd edn) Elsevier: London

Ferguson, I. (1994) 'Containing the Crisis: Crime and the Tories' *International Socialism Journal* 62

Fijnaut, C. and Marx, G.T. (1995) *Undercover: Police surveillance in comparative perspective* Kluwer: The Hague

Finn, R. and McCahill, M. (2010) 'Representing the Surveilled: Media Representations and Political Discourse in Three UK Newspapers' in Political Studies Association Conference Proceedings

Francis, M. (1996) '"Set the people free"? Conservatives and the state, 1920–1960' in M. Francis and I. Zweiniger-Bargielowska (eds) *The Conservatives and British Society, 1880–1990* University of Wales Press: Cardiff pp.58–77

Fussey, P. (2004) 'An Interrupted Transmission? Processes of CCTV implementation and the impact of human agency' *Surveillance and Society* 4(3) pp.229–56

Fyfe, N.R. and Bannister, J. (1996) 'City Watching: closed circuit television surveillance in public spaces' *Area* 28 pp.37–46

Fyfe, N.R. and Bannister, J. (1998) 'The Eyes Upon the Street: Closed-Circuit Television Surveillance and the City' in N.R. Fyfe (ed.) *Images of the Street* Routledge: London pp.254–67

Gamson, W.A. (1988) 'Political Discourse and Collective Action' *International Social Movement Research* 1 pp.219–46

Gamson, W.A. and Modigliani, A. (1989) 'Media Discourse and Public Opinion on Nuclear Power: A constructionist approach' *American Journal of Sociology* 95 pp.1–37

Gandy, O.H. (1989) 'The Surveillance Society: Information Technology and Bureaucratic Social Control' *Journal of Communication* 39(3) pp.61–76

Gellman, R. 'A General Survey of Video Surveillance Law in the United States' in S. Nouwt, B.R. de Vries and C. Prins (eds) *Reasonable Expectations of Privacy: Eleven Country Reports on Camera Surveillance and Workplace Privacy* TMC Asser Press: The Hague

Gerrard, G. Parkins, G. Cunningham, I. Jones, W. Hill, S. and Douglas, S. (2007) *National CCTV Strategy* HMSO: London

Giddens, A. (1998) *Conversations with Anthony Giddens: Making sense of modernity* Stanford University Press: California

Gill, M. Smith, P., Spriggs, A. Argomaniz, J. Allen, J. Follett, M. Jessiman, P. Kara, D. Little, R. and Swain, D. (2003) *National Evaluation of CCTV:*

Early Findings on Scheme Implementation, Effective Practice Guide Home Office Development and Practice Report, no.7, HMSO: London

Gill, M. et al. (2005) 'The Impact of CCTV: Fourteen case studies' Home Office Online Report 15/05: http://www.homeoffice.gov.uk/rds/pdfs05/rdsolr1505.pdf

Gill, M. and Spriggs, A. (2005) *Assessing the Impact of CCTV,* Home Office Research Study, no. 292, HMSO: London

Gill, M., Bryan, J. and Allen, J. (2007) 'Public Perceptions of CCTV in Residential Areas: "It Is Not As Good As We Thought It Would Be"' *International Criminal Justice Review* 17 pp.304–24

Gilling, D. (1994) 'Multi-agency Crime Prevention in Britain: The problem of combining situational and social strategies' in R.V. Clarke (ed.) *Crime Prevention Studies* Vol.3 Criminal Justice Press: New York pp.231–48

Glynn, S. and Booth, A. (1996) *Modern Britain: An economic and social history* Routledge: London and New York

Goodchild, 22 December 1958 labelled 12/730/120/ Police 1606 (Open University archive)

Goodwin Gerberich, S., Gibson, R.W., Gunderson, P.D., Melton, L.J., French, L.R., Renier, C.M., True, J.A. and Carr, W.P. (1991) 'Surveillance of Injuries in Agriculture' in Myers, M.L. (1994) *Papers and Proceedings of the Surgeon General's Conference on Agricultural Safety and Health* DIANE Publishing.

Goold, B. (2004) *CCTV and Policing: Public area surveillance and police practices in Britain* Oxford University Press: Oxford

Graham, S. (1996) 'CCTV – Big Brother or Friendly Eye in the Sky?' *Town and Country Planning* 65(2) pp.57–60

Gras, M.L. (2004) 'The Legal Regulation of CCTV in Europe' *Surveillance and Society* 2(2/3) pp.216–19

Gregory, J. (2001) 'Public Understanding of Science: Lessons from the UK experience' SciDev.net: http://www.scidev.net/global/communication/feature/public-understanding-of-science-lessons-from-the.html

Gregory, J. and Lock, S.J. (2008) 'The Evolution of "Public Understanding of Science": Public engagement as a tool of science policy in the UK' *Sociology Compass* 2(4) pp.1252–565

The Guardian 13 May 1993 'Big Brother is Here'

The Guardian 30 January 1995 'Spy Cameras Become Part of Landscape'

The Guardian 22 March 1995 'Security: Someone's Watching'

The Guardian 21 September 1995 'Surveillance: Spies on the Streets'

The Guardian 23 November 1995 '10,000 Spy Cameras for High Streets'

The Guardian 9 January 1998 'Demand for Closed Circuit TV Triggers Fear of Crime'

The Guardian 15 October 1998 'Fears for Civil Liberties: Spy camera matches faces to police files'

The Guardian 10 February 2000 'In Your Face'

The Guardian 1 December 2000 'Witness clue in Damilola murder'

The Guardian 2 December 2000 'Witness saw Boy Minutes Before his Death'
The Guardian 5 December 2000 'Caught on Camera: The last journey of Damilola Taylor'
The Guardian 30 January 2002 'Lager Toffs'
The Guardian 26 March 2002 'Police Intensify Hunt for Missing Teenager'
The Guardian 13 July 2002 'FBI to Enhance Possible Video Footage of Milly'
The Guardian 8 August 2002 'Abduction Fear for Missing Girls'
The Guardian 9 August 2002 'Captured on Camera: Film of missing girls released'
The Guardian 9 August 2002 'Police Trace Moments Girls Vanished'
The Guardian 12 August 2002 'CCTV Plan "Could Have Foiled Snatch"'
The Guardian 7 September 2002 'Big Brother: Your data trail'
The Guardian 6 January 2005 'New Arrest in Hunt for Teenager's Killer'
The Guardian 16 January 2005 'Man Held on Amy Murder'
The Guardian 25 March 2005 'Surveillance: Little brother'
The Guardian 6 May 2008 'CCTV Boom has Failed to Slash Crime, Say Police'
The Guardian 6 March 2009 'Revealed: Police databank on thousands of protestors'
The Guardian 31 October 2012 'MPs call communications data bill "honeypot for hackers and criminals"'
The Guardian 23 May 2013 'Lib Dem opposition to communications data bill "putting country at risk"'
Gurevitch, M. and Levy, M.R. (1985) (eds) *Mass Communication Review Yearbook* (5) Sage: CA
Habib, I. (2008) *Black Lives in the English Archives 1500–1677* Ashgate: Aldershot
Haggerty, K.D. and Ericson, R.V. (eds) (2006) *The New Politics of Surveillance and Visibility* University of Toronto Press: Toronto
Hamilton, P. and Hargreaves, R. (2001) *The Beautiful and the Damned: The creation of identity in nineteenth century photography* National Portrait Gallery: London
Hanley, L. (2007) *Estates: An intimate history* Granta: London
Hansard Written Answers (accessed 18/06/2008) 'Closed Circuit Television Challenge': http://hansard.millbanksystems.com/written_answers/1996/dec/17/closed-circuit-television
Heidegger, M. (1977) *The Question Concerning Technology and Other Essays* trans. W. Lovitt Harper and Row: New York
Helten, F. And Fischer, B. (2004) 'What do people think about CCTV? Findings from a Berlin Study?' Urbaneye Working Paper Series No.13
Hempel, L. and Töpfer, E. (2002) 'Inception Report' Urbaneye Working Paper Series No.1
Hier, S. and Greenberg, J. (2009) 'CCTV Surveillance and the Poverty of Media Discourse: A Content Analysis of Canadian Newspaper Coverage' *Canadian Journal of Communication* 34 pp.461–86
Hilgartner, S. (1990) 'The Dominant View of Popularization: Conceptual Problems. Political Uses' *Social Studies of Science* 20(3) pp.519–39

Hillegas, M.R. (1967) *The Future as Nightmare: H. G. Wells and the Anti-Utopians* Oxford University Press: New York
Hoey, A. (1998) 'Techno-cops: Information Technology and Law Enforcement' *International Journal of Law and Information Technology* 6(1) pp.69–90
Holden, A. (2005) *Tourism Studies and the Social Sciences* Routledge: London
Hollowell, J. (2003) 'From Commonwealth to European Integration' in J. Hollowell (ed.) *Britain since 1945* Blackwell Publishing Ltd: Oxford
Home Office (1996) 'Protecting the Public' HMSO: London
Home Office (1997) *No More Excuses: A new approach to tackling youth crime in England and Wales* HMSO: London
Home Office (2001) *Criminal Justice: The way ahead* HMSO: London
Home office (2001) *Policing a New Century: A blueprint for reform* HMSO: London
Home Office (2002) *Criminal Justice White Paper: Justice for all* HMSO: London
Home Office (2011) 'Consultation on a Code of Practice Relating to Surveillance Cameras' HMSO: London
Home Office Press Release (10 February 2011) 'National identity register destroyed as government consigns ID card scheme to history': https://www.gov.uk/government/news/national-identity-register-destroyed-as-government-consigns-id-card-scheme-to-history
Home Office (2013) National DNA Database Statistics: https://www.gov.uk/government/publications/national-dna-database-statistics
Honess, T. and Charman, E. (1992) 'Closed Circuit Television in Public Places: Its acceptability and perceived effectiveness' Police Research Group Crime Prevention Unit Series Paper no.35
Hood, J. (2003) 'Closed Circuit Television Systems: A failure in risk communication?' *Journal of Risk Research* 6(3) pp.233–51
House of Lords Select Committee (1998) 'Digital Images as Evidence'
House of Lords Select Committee (2009) 'Surveillance: Citizens and the State'
Hoyle, C. and Rose, D. (2001) 'Labour, Law and Order' *Police Quarterly* 72(1) pp.76–85
Hughes, T. (1994) 'Technological Momentum' in M.R. Smith and L. Marx (eds) *Does Technology Drive History?* MIT Press: Cambridge pp.101–14
Human Rights Act 1998
Human Rights Watch (2011) 'UK – proposed counterterrorism reforms fall short: authorities should rely on criminal prosecution to combat terrorism'
The Independent 2 October 1992 'The Labour Party in Blackpool: Crime "encouraged by Tory Values"'
The Independent 19 October 1993 'Flood of Prisoners Alarms Governors'
The Independent 31 March 2001 '300 extra CCTV Schemes Planned'
The Independent 29 June 2002 'Half CCTV Schemes do not Reduce Crime Rate'
The Independent 23 July 2005 'London is Under Attack as Never Before, And Fear Is Not a Shameful Response'

Instruction 1/2006. Author's translation. Available in Spanish at: http://www.agpd.es/portalwebAGPD/canaldocumentacion/publicaciones/common/pdfs/guia_videovigilancia.pdf

International Herald Tribune 1 June 1998 'City of London, an IRA Target, Starts to Relax'

Introna, L.D. and Nissenbaum, H. (2009) 'Facial Recognition Technology: A survey of policy and implementation issues' Centre for Catastrophe Preparedness and Response, New York University

Ironside, M. and Seifert, R. (2000) *Facing Up to Thatcherism: The history of NALGO* Oxford University Press: Oxford

'Italy – Data Protection' (2005) in *Privacy in Research Ethics and Law*: http://www.privireal.org/content/dp/italy.php

Jain, K.A. (2004) 'An introduction to biometric recognition' *IEEE Transactions on Circuits and Systems for Video Technology* 14 (1): 1–10

Kammerer, D. (2004) 'Video Surveillance in Hollywood Movies' *Surveillance & Society* 2(2/3) pp.464–73

Katz v. United States 386 U.S. 954, 1967

Keenan, M.P. and Flanagan, K. (1998) 'Trends in UK Science Policy' in P.N. Cunningham (ed.) *Science and Technology in the United Kingdom* Cartermill: UK

Klosek, J. (2007) 'The War on Privacy' Praeger: Conneticut

Koskela, H. (2000) '"The Gaze without Eyes": Video-Surveillance and the Changing Nature of Urban Space' *Progress in Human Geography* 24(2) pp.243–65

Kroener, I. (2013) 'Caught on Camera: The Media Representation of Video Surveillance in Relation to the 2005 London Underground Bombings' *Surveillance and Society* 11(1/2) pp.121–33

Kroener, I. and Neyland, D. (2012) 'New Technologies, Security and Surveillance' in K. Ball K.D. Haggerty and D. Lyon (eds) *Routledge Handbook of Surveillance Studies* Routledge: London pp.141–8

Leeuw, K. De and Bergstra, J.A. (2007) *The History of Information Security* Elsevier: Amsterdam

Lett, D., Hier, S.P. and Walby, K. (2010) 'CCTV Surveillance and the Civic Conversation: A study in public sociology' *Canadian Journal of Sociology* 35(3) pp.437–62

Ley 23/1992 de 30 de Julio, de Seguridad Privada. Author's translation. Available in Spanish at: http://noticias.juridicas.com/base_datos/Admin/l23-1992.html

Library of Parliament (2006) 'The Right to Privacy and Parliament': http://www.parl.gc.ca/information/library/PRBpubs/prb0585-e.htm

Lippman, W. (1922) *Public Opinion* Allen & Unwin: London

Llg, R.E. and Haugen, S.E. (2000) 'Earnings and Employment Trends in the 1990s' *Monthly Labour Review* 123(3)

Luhmann, N. et al. (1993) *Risk: A sociological theory* Walter de Gruyter: New York

Lyon, D. (1994) *The Electronic Eye: The rise of the surveillance society* University of Minnesota Press: Minneapolis

Lyon, D. (2001) *Surveillance Society: Monitoring everyday life* Open University Press: Buckingham

Lyon, D. (2003) *Surveillance after September 11* Polity Press: Cambridge

Lyon, D. (2007) *Surveillance Studies: An overview* Polity Press: Cambridge

Lyon, D., Haggerty, K.D. and Ball, K. (2012) 'Introducing Surveillance Studies' in K. Ball, K.D. Haggerty and D. Lyon (eds) *Routledge Handbook of Surveillance Studies* Routledge: London and New York

MacDonald, C. (1990) *Britain and the Korean War* Blackwell: Oxford

The Mail on Sunday 21 February 1993 'Major on Criminals'

The Mail on Sunday 10 July 2005 'All Three Tube Bombs'

Manwaring-White, S (1983) *The Policing Revolution: Police, technology, democracy and liberty in Britain* The Harvester Press Limited: Sussex

Marsh, I. Melville, G. and Cochrane, J. (2004) *Criminal Justice: An introduction to philosophies, theories and practice* Routledge: London

Martin, A.K., Van Brakel, R. and Bernhard, D. (2009) 'Understanding Resistance to Digital Surveillance: Towards a multi-disciplinary, multi-actor framework' *Surveillance and Society* 6(3) pp.213–32

Marwick, A. (1996) *British Society Since 1945* (3rd edn) Penguin Books: London

Marx, G.T. (1985) 'The Surveillance Society: The threat of 1984-style techniques' *The Futurist* 19/3 pp.21–6

Marx, G.T. (2002) 'What's New about the New Surveillance? Classifying for change and continuity' *Surveillance and Society* 1(1) pp.9–29

Marx, G.T. (2005) 'Surveillance and Society' in *Encyclopaedia of Social Theory* http://web.mit.edu/gtmarx/www/surandsoc.html

Mayhew, P. et al. (1979) 'Crime in Public View' Home Office Research Study No.49. HMSO: London

Mayor's Office for Policing and Crime (2013): https://www.gov.uk/government/policies/reducing-and-preventing-crime--2/supporting-pages/community-safety-partnerships

McCahill, M. (1998) 'Beyond Foucault: Towards a contemporary theory of surveillance' in C. Norris, J. Moran and G. Armstrong (eds) *Surveillance, Closed Circuit Television and Social Control* Ashgate: Aldershot pp.40–54

McCahill, M. (2002) *The Surveillance Web: The rise of visual surveillance in an English city* Willan Publishing: Devon

McCahill, M. and Norris, C. (2002) Working Paper No. 3: CCTV in Britain: http://www.urbaneye.net

McGrath, J. (2004) *Loving Big Brother: Performance, privacy and surveillance space* Routledge: London

McLuhan, M. and Fiore, Q. (1968) *War and Peace in the Global Village* McGraw-Hill: New York

Metropolitan Police (2012) FOI Request: http://www.met.police.uk/foi/pdfs/disclosure_2012/jan_2012/2011120002975.pdf

Miller, D. and Woodward, S. (2007) 'Manifesto for a Study of Denim' *Social Anthropology* 15(3) pp.335–51

The Mirror 11 July 2005 'Blair's Warning as more Bomb Atrocities are Feared'
Monahan, T. (2006) *Surveillance and Security: Technological politics and power in everyday life* Routledge: New York
Moran, J. (1998) 'A Brief Chronology of Photographic and Video Surveillance' in C. Norris, J. Moran and G. Armstrong (eds) *Surveillance, Closed Circuit Television and Social Control* Ashgate: Aldershot
Moran, J. (2005) *Reading the Everyday* Routledge: London
Mullard, M. (1995) *Policy-making in Britain* Routledge: London
Murakami Wood, D. and Webster, C. (2009) 'Living in Surveillance Societies: The normalisation of surveillance in Europe and the threat of Britain's bad example' *Journal of Contemporary European Research* 5(2) pp.259–73
Neuman, W.R., Just, M.R. and Crigler, A.N. (1992) *Common Knowledge* University of Chicago Press: Chicago
New York Police Department (25 February 2009) Press Release: http://www.nyc.gov/html/nypd/html/pr/pr_2009_005.shtml
Newburn, T. (2008) (ed.) *Handbook of Policing* (2nd edn) Taylor and Francis: Abingdon
Ney, S. and Pichler, K. (2002) 'Video Surveillance in Austria' Urbaneye Working paper Series No.7
Inspector Nichols, 8 October 1959, handwritten note (Open University archive)
Nieto, M. (1997) 'Public Video Surveillance: Is it an effective crime prevention tool?' California Research Bureau http://www.library.ca.gov/CRB/97/05/
NoCCTV 29 July 2011: http://www.nocctv.org.uk/forum/viewtopic.php?f=10&t=180
Norris, C. (2003) 'From Personal to Digital: CCTV, the Panopticon and the technological mediation of suspicion and social control' in D. Lyon (ed.) *Surveillance as Social Sorting: Privacy, risk and digital discrimination* Routledge: London pp. 249–81
Norris, C. (2007) 'The Intensification and Bifurcation of Surveillance in British Criminal Justice Policy' *European Journal on Criminal Policy and Research* 13(1/2) pp.139–58
Norris, C. (2009) 'A Review of the Increased Use of CCTV and Video-Surveillance for Crime Prevention Purposes in Europe' European Parliament, Directorate General Internal Policies of the Union
Norris, C. and Armstrong, G. (1998) 'Introduction: Power and vision' in C. Norris J. Moran and G. Armstrong (eds) *Surveillance, Closed Circuit Television and Social Control* Ashgate: Aldershot pp.3–18
Norris, C. and Armstrong, G. (1999) *The Maximum Surveillance Society – The Rise Of CCTV* Berg: Oxford
Norris, C. McCahill, M. and Wood, D. (2004) 'The Growth of CCTV: A global perspective on the international diffusion of video surveillance in publicly accessible space' *Surveillance and Society* 2(2) pp.110–35

Norris, C., McCahill, M. and Wood, D. (2004) 'Editorial: The Growth of CCTV: A global perspective on the international diffusion of video surveillance in publicly accessible space' *Surveillance and Society* 2(2/3) pp.110–35

Nouwt, S., Vries, B.R. and Loermans, R. (2005) 'Analysis of the Country Reports' in S. Nouwt, B.R. de Vries and C. Prins (eds) *Reasonable Expectations of Privacy: Eleven Country Reports on Camera Surveillance and Workplace Privacy* TMC Asser Press: The Hague

The Observer 11 October 1998 'As UK Crime Outstrips the US, a Hidden Eye is Watching'

The Observer 2 July 2000 'Nail Bomber Trapped by Fake Penpal'

The Observer 30 July 2000 'The End of Privacy'

The Observer 7 April 2002 'New Footage of Missing Teenager'

The Observer 18 August 2002 'Holly and Jessica: The Search: Thirteen days of agony until hope finally died'

Oc, T. and Tiesdell, S. (1997) 'Safer City Centres: The role of closed circuit television' in T. Oc and S. Tiesdell (eds) *Safer City Centres: Reviving the public realm* Paul Chapman: London pp.130–42

Pan, Z. and Kosicki, G. (1993) 'Framing Analysis: An approach to news discourse' *Political Communication* 10(1) pp.55–75

Parliamentary Assembly (2008) 'Video Surveillance of Public Areas' Council of Europe

Pfaffenberger, B. (1992) 'Technological Dramas' *Science, Technology and Human Values* 17(3) pp.282–312

Philips, D. (1985) 'A Just Measure of Crime, Authority Hunters and Blue Locusts: The revisionist social history of crime and the law in Britain 1780–1850' in S. Cohen and A. Scull *Social Control and the State* Basil Blackwell: Oxford

Pierre, J. (2006) *Handbook of Public Policy* Sage: London

Poster, M. (1990) *The Mode of Information* Polity Press: Cambridge

Privacy International (2003) 'Silenced – Czech Republic': http://www.privacyinternational.org/article.shtml?cmd[347]=x-347-103758

Privacy International (2004) 'The Republic of Lithuania': http://www.privacyinternational.org/article.shtml?cmd[347]=x-347-83771

Privacy International (2007) 'Kingdom of the Netherlands': http://www.privacyinternational.org/article.shtml?cmd[347]=x-347-559513

Privacy International (2007) 'Republic of Slovenia': http://www.privacyinternational.org/article.shtml?cmd[347]=x-347-559492

Privacy International (2007) 'Video Surveillance': http://www.privacyinternational.org/issues/cctv/_index.html

Protection of Freedoms Act 2012

Rank Precision Industries Ltd, cine and photographic division, to the Secretary of ACPO (12 December 1958) (Open University archive)

Ravetz, A. (2001) *Council Housing and Culture: The history of a social experiment* Routledge: London

Rawlings, P. (2002) *Policing: A short history* Willan Publishing: Cullompton

Reeve, A. (1998) 'The Panopticisation of Shopping: CCTV and leisure consumption' in C. Norris J. Moran and G. Armstrong (eds) *Surveillance, Closed Circuit Television and Social Control* Ashgate: Aldershot pp.69–87

Regulation of Investigatory Powers (Scotland) Act 2000

Regulation of Investigatory Powers Act 2000

Reiner, R. (2000) *The Politics of the Police* (3rd edn) Harvester Wheatsheaf: Hemel Hempstead

Reiner, R. (2002) 'Media Made Criminality: the Representation of Crime in the Mass Media' in M. Maguire et al. (eds) *The Oxford Handbook of Criminology Third edition* Oxford University Press: Oxford

Roberts, L.P. (2004) 'The History of Video Surveillance – from VCRs to eyes in the sky' Evaluseek Publishing

Rodman, B. (1968) 'Bentham and the Paradox of Penal Reform' *Journal of the History of Ideas* 29(2) pp.197–210

Rule, J. (1973) *Private lives, Public Surveillance* Allen Lane: London

Samuel, R. et al. (eds) (1986) *The Enemy Within: Pit villages and the miners' strike of 1984–5* Routledge: London

Savitch, H.V. (2008) *Cities in a Time of Terror: Space, territory and local resilience* M. E. Sharpe: London and New York

The Scottish Office (1996) 'Crime and Punishment' Scottish Office: Edinburgh

Sheldon, B. and Wright, P. (2010) *Policing and Technology* Learning Matters Ltd.: Exeter

Sked, A. (2003) 'The Political Parties' in J. Hollowell (ed.) *Britain since 1945* Blackwell Publishing: Oxford

Sked, A. and Cook, C. (1993) *Post-War Britain: A political history* (4th edn) Penguin Books: London

Slobogin, C. (2002) 'Camera surveillance of public places' *Mississippi Law Journal* 72 pp.213–33

Solove, D.J. (2011) *Nothing to Hide: The false tradeoff between security and privacy* Yale University Press: New Haven and London

Spencer, M. (1995) *States of Injustice: A guide to human rights and civil liberties in the European Union* Pluto Press: London

Spitzer S. (1987) 'Security and Control in Capitalist Societies: The fetishism of security and the secret thereof' in J. Lowman R. Menzies and T. Plays (eds) *Transcarceration: Essays in the sociology of social control* Gower: Aldershot

Spriggs, A. et al. (2005) 'Public Attitudes towards CCTV: Results from the pre-intervention public attitude survey carried out in areas implementing CCTV' Home Office Online Report

State Research Pamphlet No.2 (1981) 'Policing the Eighties – The Iron Fist' Independent Research Publications Ltd.: London

Statewatch Bulletin (2009): http://www.statewatch.org/news/

Staudenmaier, J. (1985) *Technology's Storytellers: Reweaving the human fabric* MIT Press: Cambridge

Stirling, A. (2008) 'Opening Up or Closing Down': Power, participation and pluralism in the social appraisal of technology' *Science, Technology and Human Values* 33(2) pp.262–94

StopWatch 9 July 2013 'HMIC report on Stop and Search – StopWatch response' http://www.stop-watch.org/news-comment/story/hmic-report-on-stop-and-search-out-today-stopwatch-response

Storch, R.D. (1980) 'Crime and Justice in 19th-Century England' *History Today* 30 pp.32–7

The Sun 3 March 1993 'Teach Kids What's Right … Then Get Tough If They Go Wrong'

The Sun 8 August 2002 'Parents' Vigil in Church'

The Sunday Times 17 July 2005 'The Web of Terror'

The Telegraph 24 August 2009 'One Crime Solved for every 1,000 CCTV Cameras, Senior Officer Claims'

The Telegraph 6 February 2012 'Talking CCTV Cameras to be Turned Off'

Thompson, K. (1998) *Moral Panics* Routledge: London

Tilley, N. (1993) 'Understanding Car Parks, Crime and CCTV: Evaluation lessons from safer cities' Police Research Group Crime Prevention Unit Series Paper, no.42

Tilley, N. (1998) 'Evaluating the Effectiveness of CCTV Schemes' in C. Norris, G. Armstrong and J. Moran (eds) *Surveillance, Closed Circuit Television and Social Control* Ashgate: Aldershot pp.139–53

The Times 2 August 1956 'Large-Scale Oxygen Plant Opened'

The Times 13 August 1956 'Safety Television for Harwell'

The Times 17 October 1956 'Industrial Television at Calder Hall'

The Times 21 November 1956 'Princess Margaret at North Staffs College'

The Times 31 December 1956 'Innovations at Manchester'

The Times 20 February 1957 'No Real Clash of Interest'

The Times 18 June 1957 'Television on the Upper Deck'

The Times 11 July 1957 '£326,520 paid at Sotheby's for Modern Masters'

The Times 3 October 1957 'New Giant Building Rising on the South Bank'

The Times 23 January 1958 'Labour's Own Television'

The Times 16 May 1958 'Growing Use of Television in the City'

The Times 26 May 1958 'Peaceful Days at Air Terminal'

The Times 12 August 1958 'City Temple Restored'

The Times 30 October 1958 'News in Brief – Operation on Television'

The Times 3 November 1958 'Private Television for Share Prices'

The Times 6 March 1959 'Television Experiment at King's Cross'

The Times 12 March 1959 'Television from the Signal Box'

The Times 30 June 1959 'Plan to Increase Public Garage Accommodation'

The Times 16 October 1959 'Middlesex Hospital Prize Giving'

The Times 16 November 1959 'Television Used In Traffic Control'

The Times 3 December 1959 'Surgery to be Televised for Students'

The Times 26 November 1964 'Novel Crime-fighting Methods Succeed'

The Times 17 February 1965 'Experimental detection by TV in the Dark'
The Times 25 June 1965 'TV in Watch for Rail Vandals'
The Times 21 August 1965 'Television 'Eye' for Mothers'
The Times 25 August 1966 'Boys and Girls Come out to Play'
The Times 4 January 1968 'TV Cameras help London Fight against Crime'
The Times 6 July 1994 'Security Cameras Zoom in on Crime'
The Times 23 November 1995 'TV Cameras in Towns to Increase'
The Times 10 January 2000 'Looking for our Fingerprints in Cyberspace'
The Times 11 January 2000 'Who's Reading your Email?'
The Times 20 January 2000 'Is Big Brother already Watching Us?'
The Times 1 July 2000 'Crucial Tip-off came 24 hours Too Late'
The Times 19 September 2000 'If You Feel that You are Being Watched … '
The Times 10 January 2002 'Where are the Police?'
The Times 9 March 2002 'Fortress Britain'
The Times 9 August 2002 'Sunday, Teatime: Holly and Jessica pulled on Manchester United shirts, spent 29 minutes on their computer, then walked into town'
The Times 21 September 2002 'At Last, Milly's Parents find a Kind of Relief'
The Times 5 January 2005 'Missing Girl sends Desperate "Help Me" Texts to Friends'
The Times 25 February 2005 'CCTV Cameras Fail to Cut Crime'
The Times 14 March 2005 'Hi-tech Rogues Gallery puts Police in Picture'
The Times 14 May 2005 'Every Move You Make'
The Times 1 July 2005 'Privacy on Parade, at a Price'
The Times 8 July 2005 'Body Scan Machines to be used on Tube Passengers'
Töpfer, E. (2003) 'Watching the Bear: Networks and islands of visual surveillance in Berlin' Urbaneye Working Paper Series No.8
Töpfer, E. (2012) 'From Privacy Protection towards Affirmative Regulation: The politics of police surveillance in Germany' in F. Björklund and O. Svenonius (eds) *Video Surveillance and Social Control in a Comparative Perspective* Routledge: New York pp.171–88
Tricomi, A.H. (1996) *Reading Tudor-Stuart Texts through Cultural Historicism* University Press of Florida: Florida
Verbeek, P. (2005) *What Things Do? Philosophical Reflections on Technology, Agency and Design* Pennsylvania State University Press: University Park
Wajcman, J. (1991) *Feminism Confronts Technology* Polity Press: Cambridge
Wakeman, R. (2003) *Themes in Modern European History since 1945* Routledge: London and New York
The Washington Post 12 October 2006 'Street Cameras are Likely to Stay'
Webb, B. and Laycock G. (1992) 'Reducing Crime on the London Underground: An evaluation of three pilot projects' HMSO: London
Webster, C.W.R. (2012) 'Surveillance as X-ray: Understanding the Surveillance State' in C.W.R. Webster G. Galdon Clavell N. Zurawski K. Boersma B. Ságvári C. Backman and C. Leleux (eds) *Living in Surveillance Societies: 'The State of Surveillance'* Proceedings of LiSS Conference 3

Weichert, T. (1998) 'Audio- und Videoüberwachung. Kontrolltechniken im öffentlichen Raum' *Bürgerrechte & Polizei* 60 pp.12–19 (Author's translation)

Weiser, M. (1991) 'The Computer for the Twenty-First Century' *Scientific American* September, 1991 pp.94–104

Welsh Campaign for Civil and Political Liberties (WCCPL) (1985) 'Striking Back'

Welsh, B. and Farrington, D. (2002) *Crime Prevention Effects of Closed Circuit Television: A systematic review*, Home Office Research Study, no. 252, HMSO: London

Welsh, B.C. and Farrington, D.P. (2002) 'Crime Prevention Effects of Closed Circuit Television: a systematic review' Home Office Research Study

Welsh, B.C. and Farrington, D.P. (2009) 'Public Area CCTV and Crime Prevention: An updated systematic review and meta-analysis' *Justice Quarterly* 26(4)

Williams, C.A. (2003) 'Police Surveillance and the Emergence of CCTV in the 1960s' in M. Gill (ed.) *CCTV* Perpetuity Press: Leicester

Williams, C.A. (2003) 'Police Surveillance and the Emergence of CCTV in the 1960s' *Crime Prevention and Community Safety* 5 pp.27–37

Williams, K.S. and Johnstone, C. (2000) 'The Politics of the Selected Gaze: Closed Circuit Television and the Policing of Public Space' *Crime, Law and Social Change* 34(2) pp.183–210

Wilsdon, J. and Willis, R. (2004) *See-through Science: Why public engagement needs to move upstream* Demos

Wrigley, C. (1997) *British Trade Unions 1945–1995* Manchester University Press: Manchester

Young, J. (1999) *The Exclusive Society: Social exclusion, crime and difference in late modernity* London: Sage

Zander, M. (1990) *The Police and Criminal Evidence Act, 1984* (2nd edn) Sweet & Maxwell: London

Electronic Resources

http://engage.barnet.gov.uk/customer-support-group/consultation-on-cctv-camera-locations/consult_view

http://www.birmingham.gov.uk/cs/Satellite?c=Page&childpagename=Highways-and-Maintenance%2FPageLayout&cid=1223092719990&pagename=BCC%2FCommon%2FWrapper%2FWrapper

http://www.datenschutz.de/recht/grundlagen/

http://www.europarl.europa.eu/charter/pdf/text_en.pdf

http://www.garanteprivacy.it/web/guest/home/docweb/-/docweb-display/docweb/1734653

http://www.gedling.gov.uk/community/communitycrimeprevention/closedcircuittelevisioncctv/

http://www.ico.org.uk/about_us/our_organisation/history

http://www.lbbd.gov.uk/CommunityPeopleAndLiving/CommunitySafety/Pages/CCTV.aspx

http://www.peterborough.gov.uk/services_a-z.aspx?ServID=26
http://www.statewatch.org/Targeted-issues/EU-USA-dp-agreement/eu-usa-dp-info-exchange-agreement.htm
http://www.trafford.gov.uk/communityandliving/communitysafety/cctv/
www.londonconnects.gov.uk/docman/digital-london/digital-bridge-presentation/download.html
www.londonconnects.gov.uk/docman/digital-london/digital-bridge-presentation/download.html
www.londonconnects.gov.uk/docman/digital-london/digital-bridge-presentation/download.html

Index

For Product Safety Concerns and Information please contact our EU representative GPSR@taylorandfrancis.com Taylor & Francis Verlag GmbH, Kaufingerstraße 24, 80331 München, Germany

Printed and bound by CPI Group (UK) Ltd, Croydon, CR0 4YY
01/05/2025
01858586-0001